FOREWORD

Technology is a critical element of any response to climate change in the energy sector. Yet today's technologies are not being deployed quickly enough, or on a sufficiently large scale, to reduce greenhouse gas emissions significantly in the near term. Long-term research and development (R&D) resources are becoming scarcer, making it more difficult to develop new technologies that can reduce emissions in the long term. Government and industry, working together, have important roles to play in turning these trends around. Their efforts must start today and be sustained over time.

These are the key messages of this report, which is a product of the IEA's Committee on Energy Research and Technology (CERT). The Committee developed the report as a background document for the meeting of IEA Energy Ministers in May 1999. It points out that maximising energy technology's contribution to reducing greenhouse gas emissions has both near-term and long-term components. Both are indispensable. The report urges Ministers to recognise that technology policies, and policies aimed at making it more expensive to emit carbon, are mutually reinforcing. Both are needed to reduce emissions and to lessen the financial burden associated with doing so.

The report brings together information and findings from several IEA and Member country reports and analyses. It reviews a large number of technologies that could prove important to reducing energy-related emissions in the near term and some of the barriers that must be overcome if they are to do so. It also previews some of the more fundamental changes in how energy services are produced and used that will be needed in the future, and some of the R&D needed to make those changes possible. It reviews areas where a government role seems essential to maximising technology's contribution to reducing emissions in the energy sector.

The Committee would like to thank all persons in IEA Member countries and the IEA Secretariat involved in preparing this report. We are particularly indebted to the many experts from the IEA collaborative research and development agreements (termed "Implementing Agreements") and the CERT's Working Parties and Expert Groups who made invaluable substantive contributions to the report. Comments on the report may be sent to Madeline Woodruff at the IEA Secretariat (tel.: [+33-1] 40.57.67.82, fax: [+33-1] 40.57.67.59, e-mail: madeline.woodruff@iea.org).

Sue Kirby
Chair
IEA Committee on Energy Research and Technology

Hans Jørgen Koch
Director
IEA Office of Energy Efficiency, Technology and R&D

**INTERNATIONAL
ENERGY AGENCY**

ENERGY TECHNOLOGY AND CLIMATE CHANGE

A Call to Action

OECD

INTERNATIONAL ENERGY AGENCY
9, rue de la Fédération,
75739 Paris Cedex 15, France

ORGANISATION FOR ECONOMIC CO-OPERATION AND DEVELOPMENT

The International Energy Agency (IEA) is an autonomous body which was established in November 1974 within the framework of the Organisation for Economic Co-operation and Development (OECD) to implement an international energy programme.

It carries out a comprehensive programme of energy co-operation among twenty four* of the OECD's twenty nine Member countries.

The basic aims of the IEA are:

■ To maintain and improve systems for coping with oil supply disruptions;

■ To promote rational energy policies in a global context through co-operative relations with non-Member countries, industry and international organisations;

■ To operate a permanent information system on the international oil market;

■ To improve the world's energy supply and demand structure by developing alternative energy sources and increasing the efficiency of energy use;

■ To assist in the integration of environmental and energy policies.

*IEA Member countries: Australia, Austria, Belgium, Canada, Denmark, Finland, France, Germany, Greece, Hungary, Ireland, Italy, Japan, Luxembourg, the Netherlands, New Zealand, Norway, Portugal, Spain, Sweden, Switzerland, Turkey, the United Kingdom, the United States. The European Commission also takes part in the work of the IEA.

Pursuant to Article 1 of the Convention signed in Paris on 14th December 1960, and which came into force on 30th September 1961, the Organisation for Economic Co-operation and Development (OECD) shall promote policies designed:

■ To achieve the highest sustainable economic growth and employment and a rising standard of living in Member countries, while maintaining financial stability, and thus to contribute to the development of the world economy;

■ To contribute to sound economic expansion in Member as well as non-Member countries in the process of economic development; and

■ To contribute to the expansion of world trade on a multilateral, non-discriminatory basis in accordance with international obligations.

The original Member countries of the OECD are Austria, Belgium, Canada, Denmark, France, Germany, Greece, Iceland, Ireland, Italy, Luxembourg, the Netherlands, Norway, Portugal, Spain, Sweden, Switzerland, Turkey, the United Kingdom and the United States. The following countries became Members subsequently through accession at the dates indicated hereafter: Japan (28th April 1964), Finland (28th January1969), Australia (7th June 1971), New Zealand (29th May 1973), Mexico (18th May 1994), the Czech Republic (21st December 1995), Hungary (7th May 1996), Poland (22nd November 1996) and the Republic of Korea (12th December 1996). The Commission of the European Communities takes part in the work of the OECD (Article 13 of the OECD Convention).

CONTENTS

4. TECHNOLOGIES AND R&D FOR THE LONG TERM: SOME PROMISING DIRECTIONS

5. MAXIMISING TECHNOLOGY'S CONTRIBUTION: OVERCOMING BARRIERS TO TECHNOLOGY ADOPTION

LIST OF TABLES

LIST OF FIGURES

EXECUTIVE SUMMARY

This report examines the role of technology in reducing greenhouse gas emissions associated with energy production and use.

Technology has a critical role to play in reducing energy-related greenhouse gas emissions in both the near term – to contribute toward meeting the Kyoto commitments – and the long term – to continue to reduce emissions, and mitigate their growth, beyond the Kyoto time frame.

In general terms, energy-related carbon emissions can be reduced in four ways:

- by using less of energy services such as heating, lighting, mobility, motor drive and industrial drying;

- by decreasing the amount of energy required to produce a unit of energy services, through the development and use of more efficient energy-supply and end-use technologies and systems;

- by switching from fossil fuels to non-fossil fuels and from higher-carbon fossil fuels to lower-carbon fossil fuels;

- by removing carbon from fuels and combustion exhaust gases and storing it.

The use of energy services has been increasing with economic growth, and this trend is unlikely to turn around. The bulk of the reductions in energy-related greenhouse gas emissions will be achieved through the other three approaches. These reductions will require strong actions on the part of governments to increase the use of efficient and cleaner technologies – through economy-wide measures such as energy taxes and emissions cap-and-trade systems, through technology- and sector-specific measures, or through a mix of measures.

Technology- and sector-specific measures are likely to be needed to overcome barriers to wider use of advanced technologies. But such measures alone are unlikely to have sufficient impact if not reinforced by price signals or other ways of encouraging investments in low-carbon technologies.

There are many technologies that are commercial or near-commercial today that could reduce greenhouse gas emissions in the near term. Examples include heating and cooling equipment in buildings, efficient conventional vehicles, combined heat and power systems for communities and industry, natural gas combined-cycle power plants, wind energy systems, biomass energy systems and photovoltaic power systems (in certain regions). Carbon sequestration can also contribute. Fuel cells for transport and stationary power generation could make significant contributions after 2010.

But today's efficient and cleaner technologies will go only part of the way toward achieving the sustained reductions in emissions that will be needed over time to reduce the concentration of greenhouse gases in the atmosphere. Fundamental changes in how energy services are produced will be needed. Only a sustained commitment to long-term R&D, and support for deployment of long-term technologies once developed, will provide the means to make substantial emissions reductions over time.

One of the most important messages of this report is accordingly that, given the dual nature of the challenge, a technology strategy for reducing greenhouse gas emissions must *start today*, but must focus *simultaneously on the short term and the long term*.

Governments have an important role to play, often in co-operation with the private sector, in removing barriers to rapid and wide deployment of clean and efficient energy technology. In particular, governments have a potentially important role in creating and stimulating markets for new technology and in supporting "technology learning". They can do this by direct investment (through their own purchasing and through subsidising purchases by others) and by setting market rules (for example, through efficiency standards or regulations that require that a certain fraction of electricity generation come from certain sources, such as renewable fuels). Investing in alternative-fuel vehicle fleets or advanced building shell components for government's own use can help stimulate the technology learning process as well as the development of supporting infrastructure. Providing guaranteed markets for advanced products or organising small purchasers into larger groups can stimulate technology advancement and use.

Intensified efforts on longer-term R&D will be needed to develop and commercialise advanced energy technologies. Governments also have a role to play in performing this work, or stimulating the private sector to perform it. As energy market reforms and increasing global competition drive private-sector R&D toward the shorter term, government's role in supporting long-term R&D becomes increasingly important. Industry has also noted the continuing need for government support for demonstration projects, which are typically high cost.

IEA Member countries' efforts to develop and implement technology strategies to reduce greenhouse gas emissions are supported by the analysis and convening powers of the Agency, as well as its ability to engage the private sector.

CHAPTER 1
INTRODUCTION: TECHNOLOGY, CLIMATE CHANGE AND THE CHALLENGE FOR GOVERNMENTS

Technology and the Climate Challenge

In 1997, governments adopted the Kyoto Protocol[1], under which they agreed to reduce their greenhouse gas emissions by the period 2008 to 2012. A large fraction of these emissions – 85 percent across Annex I countries in 1995 – arises from the production and use of energy. As a result, the commitments made in Kyoto will require significant reductions in energy-sector emissions in many countries.

The magnitude of the challenge to IEA Member countries should not be underestimated. There is a large gap between the expected energy-related emissions in a "business-as-usual" scenario – in which no additional actions are taken to reduce emissions – and those required to meet the Kyoto commitments. OECD-wide, business-as-usual CO_2 emissions from energy production and use could be 30 percent above 1990 levels in 2010 (IEA 1998b).[2] The Kyoto commitment by Annex I countries pledges them to reduce greenhouse gas emissions to an aggregate total of 5.4 percent below 1990 levels between 2008 and 2012.

To meet this challenge, IEA Member countries are considering how to reduce energy-related greenhouse gas emissions. There are four ways:

---

1. *The Kyoto Protocol to the United Nations Framework Convention on Climate Change.*
2. *This estimate is based on the* World Energy Outlook 1998 *business-as-usual case. Underlying assumptions include: 1) the share of natural gas in primary energy supply rises relative to coal and oil, 2) some additional nuclear plants are constructed, 3) the use of renewable-energy sources increases and 4) energy use rises more slowly than economic activity.*

- Use less of energy services such as heating, lighting, mobility, motor drive and industrial drying.

- Decrease the amount of energy required to produce a unit of energy service, through the development and use of more efficient energy-supply and end-use technologies and systems.

- Switch from fossil fuels to non-fossil fuels[3] and from higher-carbon fossil fuels to lower-carbon fossil fuels.

- Remove carbon from fuels and combustion exhaust gases and store it.

Cutting back on the use of energy services is difficult and unpopular. The use of energy services is increasing with economic growth and is unlikely to turn around. Some changes in how energy services are provided may be possible, such as substituting mass transit for individual travel. But the bulk of reductions in energy-related greenhouse gas emissions will be achieved through the other three approaches.

Individual IEA Member countries are currently considering the best ways to achieve these changes. They are evaluating economy-wide measures such as carbon taxes and domestic emissions cap-and-trade systems. They are also studying sector- and technology-specific measures such as building codes, efficiency standards, targets for the use of renewable energies, voluntary agreements with industry, and research and development (R&D). And they are examining a combination of economy-wide and sector- and technology-specific measures.

Amid these discussions, however, the role of technology is sometimes overlooked. *Technology- and sector-specific measures* are intended to increase the use of efficient technology, stimulate the adoption of technologies based on new fuels, and motivate the use of technologies for removal and storage of carbon from fuel or flue gases. *Economy-wide measures* stimulate the use of new technologies by influencing the extent to which the goals of energy efficiency, fuel switching, and carbon removal and storage are pursued. Economy-wide measures also influence

---

3. *The term "non-fossil fuels" as used here includes renewable sources of energy: wind, solar energy and so forth.*

how much individuals, governments and companies make use of energy services. If advanced, more efficient and cleaner technologies than those in use today are available, it will be possible for individuals, governments and companies to reduce emissions without reducing their use of energy services, although often at some cost. If the new technologies are available at competitive cost, the shift to an energy system with lower emissions will be eased. Advanced, competitive energy technologies do more than reduce emissions: They make economy-wide price effects less burdensome and help ensure that energy services can be maintained as emissions reductions are pursued.

The Kyoto Protocol has an immediate, *near-term* impact. It requires that emissions be reduced dramatically relative to current trends within roughly the next 10 years. Technology has an important role to play in meeting this challenge. But only today's commercial and near-commercial technology will contribute in this brief time frame. Recent progress has been made in developing and commercialising efficient and cleaner technologies that can be useful in the near term. They include large-scale wind turbines, community energy systems, natural-gas-fired or coal-fired combined-cycle power plants, hybrid gasoline-electric vehicles, advanced windows and lighting for buildings, and (in some regions) rooftop photovoltaic systems. A selection of such technologies is reviewed in Chapter 3.

Technology also has an important role to play over the *long term*. To lower atmospheric concentrations of greenhouse gases in the long term, further emissions reductions will be necessary beyond 2012. Expected growth in the world's population, particularly in developing countries; increased economic activity worldwide; and improvements in living standards will all serve to accelerate growth in demand for energy services. Advances in technology will be needed that can further break the link between economic growth and greenhouse gas emissions. Chapter 4 describes some of the possible advances. Particularly needed will be innovative approaches to providing energy services with very high energy efficiency and significantly reduced emissions at low cost. In some cases, fundamentally different approaches to providing energy services will be needed. The "visioning" or "foresighting" exercises under way in many IEA Member countries will help identify such approaches and the new technologies to implement them.

Technology alone is not a panacea. Wholesale shifts to particular new technologies or systems, such as electric vehicles or fuel cells or mass transit, would require wide-ranging changes in lifestyle and behaviour. Such changes are beyond the scope of this report but should be borne in mind when estimating technology's potential contribution.

The Challenge for Governments

Most observers agree that the technologies needed to meet the Kyoto commitments are available "on the shelf" today. But they will be insufficient to achieve the needed emissions reductions; both the extent and the rate of their adoption are too low. The modelling results described in Chapter 2 support this conclusion.

As discussed in Chapter 5, a number of factors retard the pace of adoption of advanced technologies. For example:

■ Equipment and facilities are normally replaced only when replacement is cheaper than repair or refurbishment. It is common business practice to adopt new technologies when existing capital equipment is replaced. Normal capital stock turnover acts to bring efficient and cleaner technology into use. But the turnover of major capital equipment is slow – lifetimes range from 8 to 10 years for personal automobiles, to 15 years for industrial equipment, to 30 years or more for power-generation plants and industrial facilities. Buildings can last 60 to 100 years or more. Regular replacement offers only limited opportunities for the deployment of new technologies.[4]

---

4. *New capital stock additions offer additional opportunities to use new technologies. In the rapidly-growing economies of the developing world, a great deal of new infrastructure and equipment will be put into place in the coming decades, offering opportunities for investments in clean and efficient energy technologies and systems in support of these countries' sustainable development goals.*

■ New technologies are often more expensive than those they replace. Through the process referred to as "technology learning", costs fall as cumulative production grows. But as long as costs remain high, private investors alone won't make the investments in learning needed to bring them down to competitive levels. Low fossil-fuel prices exacerbate the cost barrier, particularly for replacement of fossil-fuel-based technology with non-fossil-fuel alternatives.

■ The risks associated with financing and implementing new technologies can delay investments, particularly in tight economic conditions, in less than fully competitive markets or under conditions of high uncertainty.

■ Barriers other than direct costs and inherent risk – such as lack of information and inefficiencies in market organisation – can deter investment in new technology.

In light of limiting factors such as these, if energy-related CO_2 emissions are to be reduced in the near term, government *actions are needed now to increase the use of today's efficient and clean technology*. The 2008-to-2012 time frame may seem a long way off, but it will take time to design and implement new policies and measures and time for them to have an effect. Because of the extended time required for new technology to be introduced and the modest pace of capital stock turnover, many promising technologies will have little effect before 2012. Action taken now to increase their use could ensure their growing contribution soon after 2012. Immediate and sustained action will be required if technology is to play an important role in reducing emissions by 2012 and beyond.

To achieve substantial reductions in the long term, *actions are also needed to accelerate the availability of advanced technologies and ensure their wide use once available*. These actions include support for demonstration projects, for focused long-term R&D, and for efforts to develop fundamentally different approaches to providing energy services. The lead times for technology development can run to decades when all deployment stages are considered – from basic research, to development and pilot testing, and finally to commercialisation. Even the deployment process itself can be long – typically 10 to 30 years for the majority of firms or households to adopt a new technology (IEA 1997c). Actions to accelerate the development and deployment of new technologies need to start today and to continue over time.

The challenge for governments is to increase the use of clean and efficient energy technology in the near term, to ensure the timely availability of advanced technologies over the long term, and to sustain efforts to achieve both goals over time. This challenge encompasses both near-term and long-term goals – but requires that actions to meet both sets of goals start, or be intensified, now.

To meet this challenge, governments must provide leadership and, in some cases, incentives or regulations to induce shifts in "business as usual" practices. They must combine forces with other countries to find mutually advantageous opportunities for collaboration and joint action. Partnerships with the private sector will be essential. Policy initiatives need to incorporate the role of technology into the overall climate-change strategy. Measures to accelerate technology deployment and development should be built in from the outset. Given the long lead times required for technology deployment, policies and measures must be sustained through the tenure of successive governments. The recent steady shrinking of R&D investments in many IEA Member countries and their increasingly short-term orientation raises concern regarding long-term R&D capability. The adequacy of current and planned R&D investments must be evaluated carefully in light of the immense challenge they must equip IEA Member countries to meet. R&D investments must be steady and they must be maintained.

Government and private actions to meet this challenge will typically entail up-front costs. But lower operating costs will repay much of the required investment in new technology. In addition, markets for efficient and cleaner technologies may provide an economic opportunity to suppliers that could help mitigate the costs to their countries of achieving domestic emissions reductions.

Individual IEA Member countries are currently assessing the economic feasibility and impact of achieving most of their near-term reductions domestically. There are much lower-cost opportunities to reduce emissions internationally and to contribute to sustainable development objectives at the same time. Concern has been raised, however, that achieving a large fraction of the required emissions reductions internationally may not be appropriate, because emissions from IEA Member countries have contributed so heavily to the problem.

This report does not take a stand on this point. What it does argue is that technology is critical to all response options, although it is not a panacea. It points out that there is technology available today that can help, but that supporting policies and measures that accelerate technology uptake – particularly those that provide market incentives to reduce carbon emissions – are needed to maximise its contribution. It contends that that technology development can lower the ultimate cost of restraining worldwide emissions in the long term. And it urges that actions be taken today to allow technology to put the world on a lower-emissions path tomorrow.

Contents of this Report

The remainder of this report is organised as follows:

- Chapter 2 summarises what recent modelling exercises have to say about the potential contributions of new technologies and their possible role in making economy-wide policy measures such as energy taxes or carbon cap-and-trade systems less burdensome.

- Chapter 3 reviews a selection of promising technologies for reducing greenhouse gas emissions by 2008 to 2012 and shortly beyond (through 2020). It identifies specific barriers to the wider use of each technology.

- Chapter 4 describes some of the technologies and the approaches to providing energy services that could significantly reduce emissions in the longer term, provided sufficient effort is made to develop them and ensure their wide use. It also identifies some of the areas in which supporting research is needed.

- Chapter 5 reviews the barriers to wider use of efficient and clean technology and how they might be overcome.

- Chapter 6 considers the role of government in overcoming technical barriers through support for long-term R&D as well as other measures.

- Chapter 7 synthesises the key messages of the report into the elements of an energy technology strategy for reducing greenhouse gas emissions.

- Chapter 8 proposes a role for the IEA in supporting such a technology strategy and in performing additional analysis of some of the issues raised in the following pages.

A summary of this report's main messages was given to Energy Ministers when they met in May 1999. The summary is included as Appendix C.

CHAPTER 2
SOME INSIGHTS
FROM MODELLING:
TECHNOLOGY'S ROLE

Introduction

Computer scenarios for greenhouse gas emissions reduction provide some insights into the role of technology in achieving reductions and how its use affects the cost of achieving them. A group of modellers was brought together recently by the IEA to analyse a variety of such scenarios. This group includes experts from IEA Member countries and the European Commission (and is referred to here as the "IEA/EU Experts Group"). Its work complements similar studies by the IEA Implementing Agreement[5], "Energy Technology Systems Analysis Programme" (ETSAP), and by individual IEA Member countries. An example of the latter is a recent analysis by the U.S. Energy Information Administration (EIA) of several emissions reduction scenarios for the United States (U.S. DOE/EIA 1998). All of these studies have examined the potential contributions of today's energy technology, and energy technology yet to be developed, to meeting the Kyoto commitments. In some cases, they have also considered reducing emissions beyond the Kyoto time frame.

This chapter summarises some of the main findings from these efforts.

Main Findings

Some of the key messages from computer modelling efforts are reviewed below.

---

5. *"Implementing Agreements" are collaborative R&D agreements among IEA Member countries and some non-Member countries.*

- Technology improvements can reduce the ultimate cost of achieving the Kyoto commitments and of further emissions reductions beyond the Kyoto time frame.

- Additional policies and measures – including some that send price signals – will be needed to stimulate the technology improvement and adoption needed to meet the Kyoto commitments.

- Electricity generation from renewable-energy technologies is likely to increase substantially by 2020-2030 in absolute terms, but will still remain a small fraction of total generation.

- Competition between improved clean-coal technologies and natural gas technologies for power generation is a complex story.

- The potential contribution of technological advances in end-use technologies, and their rapid deployment, is significant.

Technology Improvements Can Reduce the Ultimate Cost of Emissions Reduction

Work by both the EIA and the IEA/EU Experts Group indicates that technology improvements reduce the ultimate cost of achieving the Kyoto commitments. They also reduce the cost of further emissions reductions beyond the Kyoto time frame.

For example, the EIA's analysis for the United States (U.S. DOE/EIA 1998) includes two cases designed to test the effects of both the technology improvements already embedded in a "reference case", and further technology improvements beyond the reference case. Both energy consumption and energy production technologies are addressed.

The two technology-sensitivity cases, and the reference case with which they are compared, are as follows:

- **The Reference Case** assumes that average energy-related carbon emissions are 9 percent above 1990 levels in the period 2008 to

2012. Reference assumptions about the rate of improvement in the efficiency of energy supply and end-use technologies, and the maximum market shares attained by these technologies, are embedded in this case.

- **The High-Technology Case** assumes faster development (than in the reference case) of more energy-efficient and low-carbon technologies for energy production and use. It also assumes lower costs, higher maximum market shares and higher efficiencies for these technologies.[6] These improvements are assumed to have resulted from increased R&D. They do not assume the more rapid adoption of advanced technologies that is induced by higher energy prices.

- **The Low-Technology Case** assumes static technology performance after 1998. All future choices of equipment are assumed to be made from end-use and power-generation equipment available in 1998, building shell and industrial plant efficiencies are frozen at 1998 levels, and it is assumed that no new electricity-generation technologies become available. The technology performance available to the market is kept artificially below that available in the reference case.

For all three cases, the average emissions level over the 2008-to-2012 Kyoto commitment period is the same. The only difference is the carbon price required to attain that level. The results are shown in Table 2.1.

This analysis shows that, for the United States, technology development lowers the ultimate cost of reducing emissions.

---

6. *The assumptions about technology improvements were developed for individual technologies or groups of technologies. For example, in the high-technology case, the best electric water heater available in 2015 is assumed to be 17 percent more efficient than the 1998 model, whereas in the reference case the best available technology is only 4 percent more efficient than in 1998. The water heater cost in the high-technology case is assumed to be 27 percent below the 1998 cost. (The cost assumed in the reference case is unchanged from 1998.) In the pulp and paper industry, the energy intensity of the paper-making process step in 2015 in the high-technology case is 38 percent below the intensity in 1998, whereas in the reference case the intensity is only 23 percent lower than in 1998.*

Table 2.1

Projected Carbon Prices for U.S. Technology-Sensitivity Cases

	Low-Technology Case	Reference Technology Case	High-Technology Case
Projected Carbon Price (1996 US$ per metric ton)	$243	$163	$121
Carbon Price Relative to Reference Case	+ 49%	–	– 26%

Source: U.S. DOE/EIA 1998.

The reference-case assumptions about technology development and adoption yield a carbon price of $163 per metric ton for achieving the desired emissions reduction. The low-technology case demonstrates that, without the technology improvements and market shares included in the reference case, the cost of achieving the same amount of emissions reduction would be 49 percent higher. Even the normal rates of technology improvement and adoption expected today significantly lower the cost of reducing emissions, compared with a case in which technology does not improve beyond its 1998 performance.

The even higher rates of technology improvement and adoption postulated in the high-technology case lead to further reductions in the required carbon price. In this case, the cost of achieving the emissions reduction is $121 per metric ton, 26 percent lower than in the reference case and half the cost computed for the low-technology case.

Delivered energy consumption in the high-technology case is 2 percent lower than in the reference case, even after account is taken of the "rebound effect".[7] Average energy prices, including carbon prices, are 10 percent lower. As a result, total direct expenditures on energy are 13 percent lower in the high-technology case than in the reference case.

The IEA/EU Experts Group had similar findings. It evaluated a series of scenarios involving significant cost and performance improvements in specific electricity-generation technologies and in end-use technologies, resulting primarily from enhanced R&D. It found that, given a particular emissions limit, greater technology cost reduction and greater market adoption of advanced supply and demand technologies would help in achieving the limit at lower cost.

Additional Policies and Measures – Including Some that Send Price Signals – Will Be Needed

"Baseline" or " business-as-usual" modelling results show that current trends in technology adoption and improvement, under current policies, will not in themselves enable IEA Member countries to achieve their Kyoto commitments. The rate of technology improvement observed today is too slow, and the rate of investment in the deployment of advanced technology is also too slow, to provide sufficient reductions in emissions to achieve the Kyoto commitments.

Policies and measures that speed the rate of technology improvement and adoption will be needed. Most likely, they will need to affect energy prices in order to send the appropriate signal to markets about the value of investments to reduce carbon emissions.

---

7. *Lower energy demand results in lower energy prices, which stimulate some increased demand, a result referred to as the "rebound effect". In the EIA's analysis, the high-technology case leads to lower delivered energy consumption in the industrial and transport sectors, as efficiency improvements in industrial processes and most transport modes outweigh the countervailing effects of lower energy prices. In the residential and commercial sectors, the effect of lower energy prices balances the effect of advanced technology, and consumption levels are at or near those in the reference case. In the power-generation sector, coal use is 40 percent higher than in the reference case, due to efficiency improvements and the lower carbon price. (The overall U.S. emissions level is the same for all cases.) In the low-technology case, converse trends prevail – energy demand is higher in the industrial and transport sectors and lower in the residential and commercial sectors. The EIA concludes that these trends suggest the industry and transport sectors are more sensitive to technology changes than to price changes, and the residential and commercial sectors are more sensitive to price changes.*

The need for additional policies and measures is demonstrated most clearly by the analysis conducted by the IEA/EU Experts Group, which quantified the impact on CO_2 emissions of substantial cost and performance improvements in a variety of energy supply and demand technologies. These improvements were assumed to have resulted from enhanced R&D. The group found that even these improvements would not suffice to enable IEA Member countries to meet their Kyoto commitments. Even with larger and more rapid improvements in technology performance and cost, additional policies and measures – including some that affect energy prices – will be needed. Alone, significant breakthroughs in technology performance and cost would certainly reduce emissions relative to current trends, but not by enough to meet the Kyoto commitments.

The need for a combination of improved technology and policies that send price signals is reinforced by the EIA's analysis, which suggests that the industry and transport sectors are more sensitive to technology changes than to price changes, and the residential and commercial sectors are more sensitive to price changes (see footnote 7).

Renewable Energy's Contribution Is Likely to Be Small in the Near Term, but Much Larger in the Long Term

The IEA/EU Experts Group found that, in its reference scenario, electricity generation from renewable-energy technologies (except traditional hydroelectric power) increases substantially in absolute terms through the period 2020 to 2030, but it remains a small fraction of total generation. Thus, the contribution of renewable-energy technologies to meeting the Kyoto commitments is likely to be small, though it may play a much larger role in the longer term.

The technologies expected to dominate renewable-energy developments in the reference cases are wind power, biomass (including waste) and small hydropower. But even in world regions where biomass resources are considerable, the expectations for further growth are far from spectacular.

Under a scenario of significantly improved renewable-energy technology cost and performance, the capacity of renewable-based electricity generation increases sharply. Nevertheless, the share of renewable energy in total electricity generation remains small through 2030.

Competition between Improved Clean-Coal Technologies and Natural Gas Technologies Is a Complex Story

Advances in clean-coal technology could sharply reduce emissions growth from electricity generation in developing countries. They are likely to have less impact in developed countries. They could even lead to an increase in emissions if clean-coal technology displaces natural gas (or nuclear or renewable energy) and economic carbon-sequestration technologies fail to materialise.

Advanced and more cost-effective clean-coal technologies are clearly beneficial in developing countries with domestic coal resources, because they can compete more successfully with less efficient coal technologies.

In developed countries, however, the story is different. The IEA/EU Experts Group found that, if clean coal technologies improve their performance and cost faster than other technologies, they may compete more successfully with natural gas technologies than they do under business-as-usual conditions. The result could be higher use of coal for power generation and higher emissions than for the business-as-usual case.[8] (This case assumes no restraints on emissions and therefore reflects solely the effects of technology improvement.)

Coal will continue to be used in IEA Member countries. Some IEA countries have large coal reserves; they may have regions and localities where natural gas is not available; they may desire to avoid the social

---

8. *The relative costs of natural gas and coal for specific power plants, accounting for transport costs, is obviously important to this point. Also important is the wider environmental and political impact of other technologies, such as nuclear fission, that compete with clean -coal technology. The security of natural gas supplies from politically unstable parts of the world is also relevant.*

dislocation that could accompany the phasing out of coal; and there may in the future be upward pressure on gas prices, reducing the attractiveness of gas-fired technology. In these cases, upgrading existing or new coal-fired plants to use clean-coal technologies would clearly reduce emissions relative to what they would otherwise have been. But in those parts of IEA countries where both coal and gas are available, and where there are no policies restricting emissions or associating a price with them, falling prices for clean-coal technologies may lead to additional coal-based generation compared with what would otherwise have been the case. This in time would mean higher emissions unless carbon sequestration technologies become economic and accepted as an environmentally satisfactory way to deal with CO_2. In the IEA/EU Experts Group's scenario postulating substantial improvements in clean-coal technology performance and costs, total emissions actually increase in some IEA countries relative to the reference case.

Such an outcome further reinforces the need for a combination of improved technology and policies that send price signals (see above). It also highlights the potential importance of further progress in carbon sequestration technologies.

The Potential Contribution of Advanced End-Use Technologies Is Significant

The deployment together of advanced electricity supply technologies and energy end-use technologies could contribute more to emissions reduction than either of them alone. Of the two, the electricity supply sector is often viewed as easier to tackle in terms of emissions reductions because of its large, concentrated emissions sources. When analysing the supply and end-use sectors separately, however, the IEA/EU Experts Group found that technological advances in demand technologies could provide larger reductions in emissions than further technological progress in supply technologies.

Efficient and clean end-use technologies can have important emissions-reducing effects because of their broad impact throughout the energy economy. They reduce the use of both direct fuels and electricity, thereby reducing emissions from the power-generation sector as well as from end-use sectors. For example, the IEA/EU Experts Group projects that, when advanced end-use technologies are available at lower cost than in the reference case, primary energy consumption is reduced by almost 7 percent in both the United States (by 2020) and the European Union (by 2030). Carbon emissions are reduced by 7.3 percent in the United States (by 2020) and by 8.1 percent in the European Union (by 2030).

In the most optimistic case, the IEA/EU Experts Group found that significant advances in the performance and cost of both supply and demand technologies led to stabilisation of CO_2 emissions in the European Union in 2030 at the level of 2000. In the United States, the same conditions did not stabilise emissions but they did significantly moderate them. Rapid deployment of advanced end-use technologies could make a significant contribution to emissions reduction.

Work by the ETSAP confirms that no single type of action will achieve the desired greenhouse gas emissions reduction – a variety of changes in the energy supply, conversion and consuming sectors will be needed.

Conclusion

The implications of these findings for developing technology strategies to respond to climate change concerns are:

■ The development and use of advanced energy technologies is critical to lowering the cost of reducing greenhouse gas emissions from energy production and use.

■ Measures that speed technology improvement and adoption and those that send price signals are mutually reinforcing. Both are indispensable to reducing emissions.

- Renewable-energy technologies will be important to future emissions reduction but are not a panacea in the near term.

- Cheaper clean-coal technologies will be important in reducing emissions where gas is not an alternative. But, in the absence of policies restricting emissions or associating a price with them, they may serve to increase emissions unless accompanied by carbon sequestration.

- Measures to speed technology improvement and adoption should address both supply-side and end-use technologies.

CHAPTER 3
SOME PROMISING TECHNOLOGIES TO REDUCE GREENHOUSE GAS EMISSIONS IN THE NEAR TERM[9]

Overview

IEA Member countries have a wide array of efficient and clean technologies available for reducing greenhouse gas emissions arising from energy production and use. Among these are technologies that the Committee on Energy Research and Technology (CERT) believes can reduce emissions in the decade from now to the Kyoto attainment period and in the decade that follows. A selection of these technologies is highlighted in this chapter, not to provide an exhaustive list of technologies, but rather to convey that promising technologies are at hand now, and that their accelerated introduction into the market and their further development can have meaningful benefits both for consumers and for the global environment. The chapter also describes some of the barriers to the development and deployment of these technologies in IEA Member countries.

Cautions Regarding the Scope and Context

Three important considerations must be emphasised regarding the scope and context of this chapter:

---

9. *This chapter draws heavily on the identified references as well as on contributions and comments provided by experts from the CERT as well as other experts from IEA Member countries, from the CERT's Working Parties and Expert Groups, and from the IEA collaborative activities in R&D and information provision (termed "Implementing Agreements"). Substantial contributions were received from the Implementing Agreements and CERT subsidiary bodies listed in Appendix A.*

■ **Selected Examples, Not "Winners".** Numerous technologies are available now that are likely to play a role in improving energy efficiency and reducing greenhouse gas emissions in the various sectors of the energy economy.[10] Because a long list would be of little use to policy makers, this chapter instead provides selected examples of promising technologies. Each of them is seen by the CERT as a good candidate to reduce greenhouse gas emissions in the Kyoto time frame and somewhat beyond. Clearly, these technologies are not the only ones that can do that; they are not the result of an exercise in "picking winners".

The criteria for selection of technologies considered "promising" are described below. Technologies not included on the list may still be promising for individual countries.

■ **Informed Opinion, Not Prediction.** The selected technologies have been chosen on the basis of the "informed opinion" of CERT delegates, not on a predictive mapping of technology evolution or on a modelling exercise.

■ **Need for Simultaneous Action to Achieve Short-Term and Long-Term Targets.** There is a pressing need for action now to accelerate the use of technologies available today, to advance the development of those that are almost market-ready, and to initiate, intensify and accelerate the R&D investments and other actions necessary to influence the long-term emissions trajectory.

A vigorous effort to realise short-term gains will enable IEA Member counties to take full advantage of near-term capital stock turnover and will help build momentum for a sustained effort to reduce emissions over the long term. It is clear, however, that if countries focus their efforts to reduce greenhouse gas emissions exclusively on near-term technologies, they will risk forgoing longer-term

---

10. *More exhaustive technology lists and descriptions are available in references such as Energy Technologies for the 21st Century (IEA 1997b), IEA/OECD Scoping Study: Energy and Environmental Technologies to Respond to Climate Change Concerns (IEA 1994), the European Commission's ATLAS project (EC 1997b), the other references cited in this chapter, and in the reports of the Intergovernmental Panel on Climate Change.*

actions that afford much greater opportunities to achieve sustained emissions reductions over time.

There is little hope of achieving long-term, sustained emissions reductions without the sort of R&D described in Chapter 4 and the sort of actions discussed in Chapters 5 and 6 to influence the longer-term emissions trajectory. But actions are needed *today* to begin shifting the energy system toward one with much lower emissions over the long term. These actions include investments in R&D, investments in technology learning, and measures to influence the type of equipment and infrastructure put into place as a result of the natural turnover and expansion of energy-producing and energy-using capital stock.

The technologies discussed in this chapter offer a good starting point for actions to reduce emissions in the near term, but such measures alone are not likely to avert the trend of increasing greenhouse gas concentrations in the atmosphere. An appropriate balance must be struck between near-term actions in support of near-term goals and additional, near- and long-term actions in support of the longer-term goal of reducing emissions in a sustained way over time.

Organisation

The technologies described in this chapter are grouped into four categories:

- energy efficiency – in buildings, industry and transport;
- clean power generation;
- crosscutting technologies;
- technologies for carbon sequestration.

The specific technologies described in each category are listed in Table 3.1. In subsequent sections, the following information is provided for each technology or group of technologies:

- a brief description;
- an identification of development status;

- a qualitative assessment of emissions reduction potential;
- a summary of technology-specific barriers to wide use.

All the technologies described here are considered to have the potential, if deployed rapidly, to reduce greenhouse gas emissions significantly.[11] The technologies are not discussed in priority order.

Table 3.1

Promising Technologies Discussed in Chapter 3

Energy Efficiency – Buildings	*Clean Power Generation*
• Heating and cooling technologies (for example, heat pumps and condensing gas furnaces) • Efficient lighting • Building envelope improvements: window and insulation retrofits • Building energy management systems • District heating and cooling systems • Technologies that reduce "leaking electricity" losses	• Natural-gas-fired technology • More efficient coal technologies • Renewable energy technologies: biomass and wind • Technologies for nuclear plant optimisation and life extension • Fuel cells for stationary generation
Energy Efficiency – Industry	*Crosscutting Technologies*
• Process integration • High-efficiency motors, drives and motor-driven systems • High-efficiency separation processes • Advanced end-use electro-technologies	• Combined heat and power • Advanced gas turbines • Sensors and controls • Power electronics
Energy Efficiency – Transport	*Carbon Sequestration*
• Efficient conventional vehicles • Electric and hybrid vehicles • Fuel-cell-powered vehicles • Biofuels	• Carbon dioxide separation technologies • Geologic storage of carbon dioxide

---

11. In some cases, the emissions reductions are significant in the sense that they represent a relatively large fraction of total emissions. Other cases represent significant reductions in categories of energy use (and resulting emissions) that do not constitute a large fraction of energy consumption. Added up, however, reductions in the latter category can be significant. Reductions of both types will be needed to meet emissions reductions goals, and so both are included here.

Selecting Technologies for Near-Term Attention

To select a set of promising technologies for near-term attention, the CERT developed a set of background assumptions and a set of criteria.

Assumptions

For purposes of this report, it was assumed that:

- new fossil-fuel-fired power plants built in IEA Member countries will use predominantly natural gas;

- no new nuclear plant orders are likely in IEA Member countries outside of Japan and France;

- small-scale infrastructure changes are feasible, but no major infrastructure changes can be expected in the near term.

Some of the technologies described here are not being used today, or are not used more widely, because of barriers such as high cost or lack of certain infrastructure. It was assumed that such barriers could be addressed by policies and measures in IEA Member countries.

Criteria

The technologies described in this chapter were selected after considering their potential to reduce greenhouse gas emissions; the location, time frame and cost of that reduction; and the potential applicability of the technologies across IEA Member countries.

- **Impact:** The technologies described here are those that the CERT believes can have a *significant* impact on greenhouse gas emissions arising from energy production and use. Technologies not included could also have a significant impact. Technologies judged to have marginal impact have not been included.

- **Location:** The technologies considered are those that could reduce energy-related emissions *in IEA Member countries*. The potential use of the Kyoto "flexibility mechanisms" was not considered, although many of the technologies discussed here could be used under these mechanisms. Where a technology may be viewed as especially relevant outside IEA Member countries, this is noted in the text, but it was not a factor in selecting technologies.

- **Time Frame:** The technologies considered are those that could have an effect on emissions *within the Kyoto time frame and shortly beyond* (to 2020). Effects by 2012 would contribute to meeting the Kyoto commitments. Because of the time required for market penetration, only technologies commercially available today could be expected to have much of an effect by about 2010. But a study restricted to this time frame would overlook some extremely promising technologies that could have an important effect in the subsequent decade. Therefore, the technologies included in Chapter 3 can be described as those "ready today" and those "almost ready" – the latter category requiring some additional R&D before full commercialisation.

- **Cost:** Many of the technologies discussed here are more expensive to purchase and operate than higher-carbon-emitting counterparts. This is particularly apparent under conditions of low fossil-fuel prices. The CERT did not reject promising technologies on the basis of cost, assuming that policies and measures of some kind would be put into place to address the cost differential.

- **Applicability:** There are many technologies that could reduce emissions substantially in only a few IEA Member countries. The promising technologies described in this chapter have reasonably broad (but not necessarily full) applicability across IEA countries.

Some Promising Energy-Efficiency Technologies: Buildings

Building stock turns over only very slowly, but the equipment used in residential and commercial buildings – heating and cooling equipment, lighting and so forth – has a much shorter lifetime. There is a significant opportunity to replace it with more efficient equipment and systems by 2010 and shortly beyond. Building retrofits also provide the opportunity to improve building shell components such as windows and insulation.

Heating and Cooling Technologies[12]

Description: Efficient heating, ventilation and air-conditioning (HVAC) equipment, as well as water heating and refrigeration equipment, is particularly promising for reducing greenhouse gas emissions. Two examples of particular interest are high-efficiency heat pumps and condensing gas furnaces.

■ **Heat Pumps:** A heat pump absorbs heat at a low temperature from an external heat source (or from internal exhaust air) and delivers it at a higher temperature to the heating system of a building. Alternatively, it may function as a space-cooling unit, absorbing heat and rejecting it outside the building. Heat pumps may be classified as air-, water- or ground-source, depending on the external heat source, and may transfer heat to internal air or water. The vast majority of heat pumps operate on the vapour compression cycle, driven by an electric motor. A growing minority is driven by an internal combustion engine or employs the absorption principle and uses gas or waste heat as the driving energy. Electric heat pumps typically consume about one-fourth to one-half as much electricity for heating as electric resistance-based systems. They can reduce total energy consumption for heating by as much as 50 percent compared with fossil-fuel-fired boilers, and further improvements are possible.

■ **Condensing Gas Furnaces:** Gas-fired furnaces incorporating "condensing" technology use a secondary heat exchanger to recover

---

12. *Sources for this section are Bouma 1994, Broderick and Moore 2000, GRI 1996, IEA 1997b, IEA 2000I, WEC 1995, comments from the IEA Implementing Agreement on Heat Pumping Technologies and comments from IEA Member countries.*

the latent heat of water in the combustion exhaust gases. Condensing releases an additional 10 percent to 20 percent of the heat available in the products of combustion, enabling condensing gas furnaces to achieve efficiencies of 90 to 97 percent. Condensing technology can also be used in integrated space and water heaters with dramatic efficiency gains.

Development Status: Efficient electric heat pumps are on the market today, as are high-efficiency gas-fired absorption heat pumps and condensing gas furnaces. Heat pumps are also available for applications other than space heating and cooling. For example, state-of-the-art electric heat pump water heaters, which extract heat from ambient air, exhaust air, or circulating water and transfer it to water in the storage tank, have unit energy consumption as much as 70 percent lower than that of the average water heater stock. But they are up to five times more expensive than electric resistance water heaters. Heat pumps still have significant potential for improvement, while efficiency improvements in condensing gas furnaces are limited by the higher heating value of gas.

Emissions Reduction Potential: The technical potential for improvements in the efficiency of heating and cooling is large. Installation of state-of-the-art technology when equipment is replaced could significantly reduce building energy consumption across IEA Member countries. In new buildings, optimised system design offers the possibility of further savings, though these will be realised only at the rate of expansion or replacement of the building stock.

Barriers to Wider Use: A particular barrier to equipment improvements in existing and new buildings is the fact that savings from energy efficiency do not always accrue to those who make the initial investment decisions: architects, real estate developers, landlords and so forth. The cost savings that would accrue from using efficient technology are not sufficient to induce building owners who would benefit from such investments to replace equipment before it fails. Even when they are purchasing new or replacement equipment, transaction costs (the time and money to investigate all alternatives, for example) can deter investors from purchasing efficient equipment.

Higher initial equipment costs are also a deterrent. Condensing furnaces are more expensive than their non-condensing counterparts. Less expensive corrosion-resistant materials would help reduce costs. Breakthroughs are needed in reducing the cost of heat pumps, especially for heating-only markets in Europe. Breakthroughs are also needed in developing the better working fluids (fluids with better specific heat, fluids enabling fast absorption/extraction of heat, and fluids that are non-corrosive and non-freezing) that could radically increase energy efficiency. The lack of a residential air-conditioning market in Europe is also a barrier to wider use of heat pumps. For cooling, heat pumps have no competitors, but for heating, all fossil-fuel-fired boilers are competitors.

Slow capital stock turnover is another barrier to improving building equipment. Although heating and cooling systems turn over more quickly than the buildings they serve, they can still last 20 years or more, which limits the rate at which more efficient technology can penetrate the market.

Efficient Lighting[13]

Description: Efficient lighting technologies offer the potential for near-term emissions reduction. State-of-the-art technology includes electronic ballasts and compact fluorescent lamps (CFLs), high-efficiency sodium discharge lamps, lamp-based torchieres (up-lights) to replace halogen torchieres, and measures to increase lighting-system efficiency.

Development Status: Electronic ballasts and CFLs are established technology. Several high-intensity discharge lamps are available, although smaller and low-wattage lamps are being developed. Sulphur

---

13. *Sources for this section are IEA 1994, IEA 1997c, IWG 1997, U.S. DOE 1995 and U.S. DOE 1997b.*

lamps have recently been developed that can replace conventional high-intensity discharge lamps in many commercial applications. A CFL-based replacement for inefficient, high-temperature halogen torchieres has recently been developed that uses 75 percent less power, lasts longer and eliminates a fire hazard. Measures to increase the efficiency of lighting systems – controls and dimmable ballasts, time-based and occupancy- or daylight-linked controls, and efficient ballasts – are commercially available.

Emissions Reduction Potential: Efficient lighting technology can meaningfully reduce energy use. Moreover, unlike other parts of the building infrastructure, most lighting system components are replaced relatively quickly (within 10 years) and thus provide opportunities to introduce more efficient technologies on a regular basis. For example, a U.S. study projects that with development and intelligent use of more efficient lighting technologies and design, lighting energy use in the United States could be reduced by more than 50 percent by 2020, with equal or improved health, comfort and productivity.

In new buildings, designs that make maximum use of natural light and efficient lighting systems offer the possibility of further savings over time.

Barriers to Wider Use: The barriers noted for efficient building equipment also apply to lighting technologies. A barrier specific to lighting is the need to respond to very individualised consumer demands. Different types of light are needed for different activities and spaces, and lighting choices reflect aesthetic considerations as well as task needs. In addition, the lack of dedicated fixtures for compact fluorescent lights limits their application and effectiveness. There have also been "chicken and egg" problems with distributing new lighting products. Because the market for a new product is small, retailers may not offer it for sale or keep large enough stocks to allow consumers to find easily the version that suits them.

Building Envelope Improvements: Window and Insulation Retrofits[14]

Description: Building envelope improvements using advanced windows and insulation hold promise for reducing emissions in the near term. *Windows* can strongly influence a building's overall energy performance. Low-emissivity coatings reduce the transfer of heat radiation from the inside of the building to the outside. Gas-fill windows, in which the gap between multiple glazing layers is filled with low-conductivity gas, also reduce heat loss.

Proper *insulation* reduces heat loss in cold weather, keeps excess heat out in hot weather and generally helps maintain a comfortable indoor environment. Traditionally, insulation has consisted of lightweight fibrous or cellular materials with pockets of air or gas, such as glass fibre, mineral wool and expanded plastics. Recent innovations have occurred in transparent and dynamic insulation materials and in phase-change and crystal structure-change materials that can be used indoors for passive solar storage.

Development Status: There have been spectacular improvements in window thermal resistance. Low-emissivity coatings and gas-fill window technologies are technologically mature. Other advanced technologies are now commercially available, such as windows with selective coatings that reduce infrared transmittance without reducing visible transmittance. The best windows on the market insulate three times as well as their double-glazed predecessors.

Building-insulation performance has improved by a factor of two to three over the past 25 years. Superinsulations that insulate at least three times as well as today's technology will soon be available for niche markets – vacuum-powder-filled, gas-filled and vacuum-fibre-filled panels; structurally-reinforced beaded vacuum panels; and switchable evacuated panels with insulating values of more than four times those of the best currently available materials.

---

14. Sources for this section are EC 1997b, PCAST 1997, U.S. DOE 1997b and WEC 1995.

Emissions Reduction Potential: Window and insulation retrofits could reduce building-related emissions significantly, particularly in parts of Europe where many homes are poorly insulated. There are opportunities in North America as well, where 20 percent of residences are poorly insulated. Approximately 40 percent of new window sales in the United States are of advanced types (low-emissivity and gas-filled). In the United States, an estimated 20 percent of residential heating and cooling energy use is associated with losses through windows. A complete change of the stock to the most cost-effective, energy-saving window systems could reduce energy losses through windows by two-thirds.

System-level benefits result from interactions among technologies. Better windows and insulation reduce the need for heating and cooling. In new homes, the savings can be more dramatic, as when passive solar design can be used and energy-producing and energy-using equipment can be optimised for the entire building.

Barriers to Wider Use: In addition to the barriers to improving building equipment energy efficiency mentioned earlier, the slow rate of building retrofit is a major barrier to realising these improvements. In the case of insulation, the barrier is the cost of adding insulation as compared with doing nothing, because low energy prices offer little incentive to undertake the expensive process of insulation retrofitting. Windows are not usually replaced to save energy but rather for aesthetic reasons or to reduce noise.

Building Energy Management Systems[15]

Description: Building energy management and control systems are promising technologies for near-term emissions reduction. Such systems automatically regulate the operation of HVAC, lighting and other systems in buildings. These systems range from simple point-of-use timers to complex microprocessor-based systems that can minimise

---

15. *Sources for this section are EC 1997b, U.S. DOE 1997b and WEC 1995.*

unnecessary equipment operation and fulfil other functions such as economiser cycling or varying supply air or water temperatures.

Development Status: Some systems, such as timer-controlled water heaters, are relatively simple and have been in use for many years. Others are highly complex and can produce levels of automation that make possible significant energy savings. The technology is commercially available.

Emissions Reduction Potential: Computerised energy management systems typically provide a 10 percent to 20 percent energy savings, although savings of 30 percent or more are possible in existing commercial buildings, even in many thought to be working properly now. There are also system benefits; energy management systems that turn off lights in unoccupied spaces can reduce the lighting heat load, and thus the need for air conditioning. In addition, energy management systems could be more effective when used in conjunction with energy-storage and waste-heat-reclamation systems.

Barriers to Wider Use: In addition to the barriers mentioned earlier for HVAC equipment and building envelope improvements, high installation costs can deter users from installing energy management systems on a retrofit basis. Lack of user friendliness in some complex systems has been a barrier to their use, although more advanced systems can also be easier to use than simpler ones, because they require less overall energy management effort.

District Heating and Cooling Systems[16]

Description: Systems for heating or cooling communities – district heating and cooling (DH&C) systems, also referred to as community energy systems – are promising for near-term emissions reduction. These systems are used primarily for heating buildings in the winter,

---

16. *Sources for this section are contributions from the Implementing Agreement on District Heating and Cooling, WEC 1995, comments from IEA Member countries, and information published by Euroheat & Power and the European Marketing Group, District Heating and Cooling (http://www.eu-dhc.org).*

but are also used to provide air conditioning year-round. District heating systems transport heat, in the form of hot water or steam, from a central plant to buildings via an underground, insulated pipeline. The network can range in length from several hundred metres to many kilometres.

These systems are used across Europe, North America and Japan, but primarily in Northern, Central and Eastern Europe. Extensive district heating systems are found in a number of European countries. In Finland and Denmark, for example, district heating systems account for 50 percent of the space-heating market. About 80 percent of the world's DH&C is located in Central and Eastern Europe; such systems are also used widely in Korea and China.

DH&C systems can use a wide variety of energy sources, including industrial waste heat and condenser heat from numerous types of thermal power generation.[17] In some countries (for example, in Denmark, Finland and Germany), the share of combined heat and power (CHP) in district heating production is as high as 62 to 70 percent.

Development Status: The technology is fairly well developed, although there is considerable potential for improved technical efficiency and reduced costs. Technical efficiency improvements are possible through better insulation of the pipes that carry heat to and among buildings, improved methods for operation and status control, and methods for remote supervision of consumer installations. Harmonised design and installation methods for distribution systems are also needed.

Emissions Reduction Potential: The potential energy savings are high in individual applications, both new and retrofit. Savings are greatest when district energy systems make use of waste heat from

---

17. *The temperature of condenser heat is too low for district heating applications, which require heat at a temperature of 60°C to 120°C. The condenser heat must be upgraded via heat extraction from the steam turbine, which reduces electricity generation by about 10% to 20% of the heat energy or requires additional fuel.*

thermal power generation. Prospects for growth are good in several EU countries, such as the Netherlands and Austria. The potential in North America may be increasing as interest in independent power production grows, leading to new gas-fired community cogeneration (CHP systems). Refurbishment of existing DH&C systems in Central and Eastern Europe, which have deteriorated and do not yet incorporate state-of-the-art technology, could offer considerable emissions reductions in those countries.

Barriers to Wider Use: The need for extensive investments in distribution system infrastructure is the largest barrier, particularly in view of the fact that DH&C systems are more productive if they can use waste heat from other energy-using processes, such as electricity production and industrial processes. For new or retrofit applications, putting the needed infrastructure into place involves the co-operation of various parties, including building owners, local public authorities and utilities or industrial companies. Lack of information on the benefits of DH&C systems is also a barrier to their wider use.

Technologies that Reduce "Leaking Electricity" Losses[18]

Description: "Leaking electricity" can give rise to significant greenhouse gas emissions, depending on how the electricity is produced. Home and office appliances and equipment – televisions, videocassette recorders, audio equipment, answering machines, microwave ovens, office equipment and other electronic devices – continue to use electricity while in stand-by mode or turned off. This "leaking electricity" represents as much as 10 percent of residential-sector electricity use in IEA Member countries. Significant electricity waste also occurs in the commercial sector. If all other sectors are included, leaking electricity is responsible for as much as 1 percent of IEA Member countries' CO_2 emissions. "Low-leakage" technologies can reduce such emissions.

---

18. *The source for this section is the IEA Secretariat.*

Development Status: Technologies are available today that, when installed in equipment at the time of manufacture, can limit electricity leakage to one watt for a wide range of electronic devices. For example, a Japanese company is now producing fax machines that "leak" less than one watt of electricity. Technologies are also available today to retrofit many categories of appliances and equipment (computers, fax machines, photocopiers, printers and televisions) to reduce "leaking electricity" losses.

Emissions Reduction Potential: Savings of 75 percent of current "leaking electricity" losses are technically feasible and cost-effective for new equipment, with no sacrifice of consumer features.

Barriers to Wider Use: Lack of consumer awareness of "leaking electricity" is a barrier to retrofit applications, as are low energy prices that may not stimulate demand to reduce the losses even if consumers are aware of them. In addition, many "leaking" appliances are traded internationally, so the small additional cost to install "low-leakage" technology is viewed as harmful to the international competitiveness of products. International appliance sales mean that standards or targets in individual countries may affect only a fraction of the new appliances sold in those countries.

Some Promising Energy-Efficiency Technologies: Industry

Numerous efficient technologies could reduce energy intensity and emissions in industry. Many of them are specific to individual industries, such as pulp and paper manufacturing, chemical processing or glassmaking. Others are common to several industries. The technologies described in this section fall into the latter category.

Process Integration[19]

Description: Process integration is the term used for a collection of strategies, methods and tools that focus on efficient use of resources (energy, raw materials, water and capital) on a systems level: total process, industrial site, region. The best-known and most widely applied process integration method is pinch analysis. It was originally developed to facilitate optimal heat recovery between heat sources and heat sinks, sometimes referred to as "energy cascading", which leads to significant energy savings. Powerful graphical diagrams can be used to make decisions about heat pumps, integration of distillation columns, the use of back-pressure turbines and so forth.

Development Status: Process integration applies to most sectors in the process industries (petroleum refining, chemical manufacturing, food and beverage production, and so forth). Pinch analysis in particular is an established and widely used technology for continuous processes. The scope of process integration methods has broadened considerably since the early 1980s when the emphasis was on heat recovery. Today, process integration can also be used for heat and power systems, utility systems, distillation systems, reactor systems, and even water management and wastewater treatment systems. Process integration is also expected to move into batch processing of such products as pharmaceuticals, resins and dyes.

Emissions Reduction Potential: The main applications of core process integration for energy savings, and thus also emissions reduction, have been in the chemical, petrochemical and refining industries. The methods have also been applied in the pulp-and-paper and food-and-drink industries. Projects have reduced energy consumption by about 10 to 40 percent. In new plant designs, even larger savings are possible in some industry branches.

---

19. Sources for this section are contributions from the Implementing Agreement on Process Integration, EC 1997b, IWG 1997, WEC 1995 and comments from IEA Member countries.

Barriers to Wider Use: The main barriers to further use of process integration are lack of awareness about the value, benefits and broadened scope of process integration and the concern that process integration may affect plant operation, and thus production, in a negative way.[20] Low energy prices also act as a barrier to use of process integration strategies. When energy prices are low, retrofit projects suffer from long payback times, unless energy-saving projects can be combined with process modifications. (Some retrofit projects with payback times from 6 months to 3 years have been reported.)

High-Efficiency Motors, Drives and Motor-Driven Systems[21]

Description: High-efficiency motors, drives and motor-driven systems hold significant potential for reducing emissions in the near term. Energy-efficiency opportunities in these systems derive not so much from the replacement of older motors with high-efficiency models as from energy-conscious design throughout the system. The system includes power supply lines, controls, motor feed cables, the electric motor, the drive and transmission system, and the driven load. Each of these system elements may present a real opportunity to conserve energy.

Development Status: Efficient technologies are available on the market. Power electronic switching devices and micro-electronics have made electronic adjustable speed drives increasingly popular and have brought down their price. Adjustable speed drives are commercially available in a large variety of designs.

Emissions Reduction Potential: Opportunities to improve motor-driven systems are large, especially in motor-driven pump and

---

20. *Sometimes this concern is justified, and more R&D is needed to overcome this problem. In a new design, heat integration reduces start-up and shut-down capabilities as well as general controllability. Process integration can, however, also be used to overcome such operational problems. Most plants today are heavily integrated, and it has been demonstrated that the flowsheet structure (process topology) is more important for good operation than the control system. Process integration can be used to identify good structures that also keep energy consumption low.*

21. *Sources for this section are EC 1997b, IWG 1997 and comments from IEA Member countries.*

compressor systems. Motor over-sizing for the driven load is common practice and provides a significant opportunity for efficiency improvement. Drive-transmission efficiency ranges from below 50 percent to well over 90 percent, suggesting substantial potential for improvement in some systems. Large efficiency gains could be achieved by better matching of system requirements through a variety of techniques, speed control with adjustable speed drives, equipment resizing and alternative fluid system control strategies. Adjustable speed drives offer the single largest opportunity to increase a drive system's energy efficiency.

Barriers to Wider Use: Low rates of capital stock turnover in motors, motor drive systems and the loads driven by motors limit the penetration of high-efficiency motor systems. Higher initial cost can be a barrier to adjustable speed drives. Lack of market awareness of the benefits to be gained from optimising the entire motor system is another barrier. In addition, the need to tailor improvements to specific systems makes broad implementation difficult.

High-Efficiency Separation Processes[22]

Description: High-efficiency separation processes, such as membrane processes, freeze crystallisation and better system controls, have significant potential to reduce emissions attributable to industry in the near term. Industrial separations recover, isolate and purify products of virtually every industrial process. Today's separation processes include distillation, extraction, drying, absorption, adsorption, crystallisation, membrane-based technologies and stripping. Such processes account for a large share of energy consumption in industry – for example, they account for nearly 43 percent of the total annual energy consumption in the U.S. chemical process industries. Improvements in this area are applicable across a wide range of industries. Distillation is a dominant

22. *Sources for this section are IEA 1997b, EC 1997b, IWG 1997, U.S. DOE 1997b, WEC 1995 and comments from IEA Member countries.*

separations process for product recovery and purification. A drawback to distillation, however, is its low efficiency.

Development Status: Large improvements have already been made in the efficiency of distillation processes themselves, with major energy savings and improvements in yield. But there is still a great deal of room for their use. Most of the distillation columns in the U.S. chemical industry, for example, do not yet take advantage of the most efficient equipment.

In freeze/melt crystallisation, the thermal energy of distillation is replaced by a cooling requirement. Melt crystallisation is still in the development phase but freeze crystallisation has already been used in the food and dairy industries.

In membrane processes, the membrane provides a physical barrier through which materials pass selectively. Membrane separation processes are already in the marketplace in the production of both commodity and specialty products, particularly in the materials and chemical-processing industries. But they could be much more widely applied. Of the membrane processes, reverse osmosis and pervaporation are seen as having the greatest potential. The use of membrane processes with distillation technology in hybrid systems has only just reached the demonstration phase.

Emissions Reduction Potential: Replacing many of the current distillation and evaporation processes with membranes and crystallisation could save from 5 to 30 percent of the energy used in separation processes.

Improvements in process control could increase the efficiency of existing distillation processes. For example, distillation columns in the U.S. refining and chemical industries commonly use from 30 to 50 percent more energy than is necessary to meet product specifications. It has been estimated that applying better column controls could reduce distillation energy consumption by an overall average of 15 percent.

Barriers to Wider Use: The chief barriers to deploying advanced separation technologies are likely to be the capital expenditures required for any substantial process modification. Another major barrier is the need for advances in the understanding of basic processes (particularly for freeze/melt crystallisation). Better materials, such as more selective membranes and membranes that can stand up to harsh environments, are also needed.

Advanced End-Use Electro-technologies[23]

Description: Advanced electro-technologies for industrial end-use applications hold promise for near-term emissions reduction. These technologies can replace many fossil-fuel-based combustion processes in industry. Examples include infrared heating, drying and paint curing; ultra-violet curing; radio-frequency and microwave heating and drying; electron beam processing for metal welding and hardening; induction heating; laser-based technologies; and industrial heat pumps. Electro-technologies in general promise less energy use, less material waste, less pollution and better product quality. They can also provide greater compatibility with advanced sensors and controls, computer controls, and decentralised manufacturing operations.

Development Status: Many electro-technologies are available today and in use in the process industries.

Emissions Reduction Potential: Because of their high efficiency at the point of use, electro-technologies can reduce CO_2 emissions in many applications, even after account is taken of the fuel used in generating electricity. At one automobile manufacturing plant, the replacement of gas-fired ovens for paint drying with infrared paint curing reduced annual CO_2 emissions by 42 to 81 percent (depending on the assumptions for the electricity fuel supply). It also decreased energy costs by 93 percent, reduced the size of the system by

---

23. *Sources for this section are IEA 1998c and comments from IEA Member countries.*

700 square metres, reduced labour costs, and increased productivity by 50 percent (IEA 1998c).

Barriers to Wider Use: Lack of knowledge of or familiarity with electro-technologies may impede some use. The low rate of industrial process equipment replacement and refurbishment also constrains the market penetration of electro-technologies. Initial capital costs for retrofit applications combined with low energy prices are a general barrier to energy-saving retrofits.

Some Promising Energy-Efficiency Technologies: Transport

Vehicles run on the modern internal combustion engine have made significant technological advances during the past three decades, yet the potential for technology to reduce further still the environmental impact of conventional vehicles is far from exhausted. Substantial reductions can be expected from highly efficient conventional vehicles as well as from advanced vehicles.

Efficient Conventional Vehicles[24]

Description: A wide variety of technologies can improve the fuel efficiency of "conventional vehicles". They include lean-burn combustion, direct-injection diesel engines, turbo compressors and inter-cooling, two-stroke engines, multi-valve heads, variable-intake valve control, advanced electronics and exhaust monitoring, four- and five- speed automatic transmissions, reduced accessory drive, lightweight materials, aerodynamic design, and better lubricants. Applications of these technologies do not cause a fundamental change in the "conventional" character of the vehicle. It will still run on an

24. *Sources for this section are EC 1997b, IEA 1997b, PCAST 1997, U.S. DOE 1997b and the IEA Secretariat.*

internal combustion engine using a spark (gasoline-fuelled) or compression (diesel-fuelled) ignition cycle and will still use a conventional drive train and a conventional vehicle configuration.

Development Status: The motor-vehicle industry has adopted many of these technologies in its vehicles, thereby gaining more fuel efficiency. Despite impressive progress in adopting fuel-efficient technologies, motor vehicle fuel efficiency has not improved significantly. The improved technical efficiency has largely been offset by increases in vehicle size, weight and performance.

Emissions Reduction Potential: While many fuel-efficiency technologies are relatively common, there is an opportunity to expand further the application of the most widely used technologies (such as four- speed transmissions and multi-valve heads). It is also possible to increase the application of technologies that are not widely used, such as two-stroke engines, lean-burn combustion and advanced structural materials. In the United States, diesel engines could provide significant savings, especially in the increasingly popular "sport-utility" vehicles.

The Intergovernmental Panel on Climate Change estimates that energy-intensity reductions in light-duty vehicles that would give users a payback in fuel savings within three to four years could reduce their greenhouse gas emissions relative to projected levels in 2020 by 10 to 25 percent (IPCC 1996).

Barriers to Wider Use: It is difficult to use lean-burn gasoline, diesel or two-stroke engines while meeting the most stringent emissions standards. For example, the NO_x emissions standards that are likely to apply to 2004 model-year cars in the United States could preclude the use of both lean-burn combustion and diesel engines in light-duty vehicles. An efficient NO_x reduction catalyst is needed, as are means to reduce particulate emissions.

A broader barrier to improved fuel efficiency is a continued trend to use technical fuel-efficiency improvements for performance and luxury instead of improved fuel economy. In the United States, low gasoline prices have discouraged any serious consumer interest in fuel efficiency. Even in Asia and Europe, with much higher prices, consumer purchases are still trending toward larger, more powerful vehicles. While continued progress is expected in smaller, more efficient engines with more power, and in the further market penetration of electronic injection control, improved transmission efficiency, advanced materials and other technologies, it is unclear how much these improvements will actually lower fuel consumption.

In the absence of clear policies to reduce fuel consumption in the automobile sector, the automobile industry will continue to produce, and consumers will continue to demand and buy, relatively large and powerful vehicles. Manufacturers currently have tremendous incentives to build large, and thereby profitable, automobiles and trucks, and to wring as much production from current technology as possible. Voluntary agreements in the United States and the European Union, and regulation according to "top-runner" levels (targeted for 2010) in Japan, are examples of policies that are attempting to counteract these incentives.

Electric and Hybrid Vehicles[25]

Description: Hybrid gasoline-electric power trains for vehicles promise large emissions reductions in the near term. Electric vehicles use electric motors and batteries instead of internal combustion engines and fuel. Hybrid vehicles also use electric motors but rely on small internal combustion engines to provide electrical power. Therefore, the hybrid vehicle is still powered by gasoline or diesel fuel (or some alternative, such as natural gas, methanol or ethanol), but due to the system efficiencies of such an arrangement, relatively high fuel efficiency is achieved. The battery is recharged by the engine, so no external recharging is required.

Development Status: Both electric and hybrid vehicles are commercially available. Electric vehicles, because of limitations

---

25. *Sources for this section are EC 1997b, U.S. DOE 1997b and the IEA Secretariat.*

imposed on their performance and range by available battery technology, have relatively limited consumer appeal. In parts of the United States where their use may be required under existing environmental regulations – primarily the "zero-emissions" regulations in California, New York and Massachusetts – commercial development of electric vehicles has been accelerated. Hybrid vehicles have a much broader consumer appeal and have reached a relatively advanced stage of development, as evidenced by the Honda Insight, which is available commercially, and the Toyota Prius, which is commercially available in Japan and is expected to be available soon in Europe and the United States.

Emissions Reduction Potential: Hybrid vehicles can reduce emissions sharply. Transport-sector emissions are forecasted to grow more than any other sector, largely as a result of increasing motor vehicle travel. Hybrid vehicles offer the potential to reduce carbon emissions by 50 to 65 percent compared with conventional vehicles and therefore could be an extremely important technology to reduce CO_2 emissions after 2010. Within the Kyoto time frame, hybrid vehicles are expected to have relatively little opportunity to become a substantial fraction of the total motor vehicle fleet.

The emissions reduction potential of electric vehicles is highly dependent on the carbon-intensity of the electricity used to charge them. Nonetheless, the savings potential is moderately positive to high. Large savings will require further technological progress and the development of supporting infrastructure.

Barriers to Wider Use: Continued technology advances are expected by the motor vehicle industry, which often works in concert with governments on R&D programmes. The primary barrier to the widespread use of hybrid vehicles is likely to be cost. Most experts do not expect that, within the next 10 years, hybrid vehicles fully equivalent to their conventional-engine counterparts can be produced at an equivalent cost.

A large-scale expansion of electric vehicle use is dependent upon the provision of an easily accessed power supply network that allows rapid recharging. Moreover, electric vehicles are likely to be even more costly than hybrid vehicles because of the battery requirements. Fundamental breakthroughs in battery performance, cost and technology are needed if electric vehicles are to avoid being limited to applications with limited range and performance requirements, such as city driving. Therefore, electric vehicles appear to have a much smaller overall potential to reduce CO_2 emissions than do hybrid vehicles.

Lack of demand for these technologies is also a significant barrier. It keeps the "technology learning" process from taking place, which would reduce costs (see Chapter 5).

Fuel-Cell-Powered Vehicles[26,27]

Description: Fuel-cell-powered vehicles hold the potential for reducing transport emissions enormously in the decade after 2010. Fuel cells are electro-chemical devices that convert the chemical energy in fuels to electrical energy directly, without combustion, with high electrical efficiency and low pollutant emissions. They are similar in principle to primary batteries, except that the fuel and oxidant are stored externally, enabling them to continue operating as long as fuel and oxidant (oxygen or air) are supplied. The power system also includes a fuel processor and a power conditioner. The fuel processor converts fuels, such as natural gas, methanol, gasoline or bio-ethanol, into the hydrogen-rich fuel required by the fuel cell. With hydrogen as its fuel, the only emission stream from a fuel cell is water. When a fuel cell uses methanol or hydrocarbons as its fuel, reforming them to obtain hydrogen will produce CO_2 and other pollutants as byproducts.

For the immediate future, proton-exchange membrane fuel cells (PEMFCs) (also called solid polymer fuel cells) appear to be the clear

---

26. *Sources for this section are contributions from the Implementing Agreement on Advanced Fuel Cells, EC 1997b, IWG 1997 and comments from IEA Member countries.*
27. *Fuel cells for stationary power generation are addressed later in this chapter.*

choice among fuel-cell technologies for light-duty vehicles, because they operate at moderate temperatures and have improved rapidly in power density and decreased in cost.

Development Status: The introduction of fuel cells into mass-market vehicles appears likely to occur only after 2005. DaimlerChrysler and Toyota have already announced demonstration vehicles with the same performance and range as conventional cars powered by internal combustion engines. In fact, all the world's leading car makers are now actively involved in evaluating fuel cells for cars. DaimlerChrysler has already demonstrated a methanol-powered fuel-cell car and is heading an ambitious programme aimed at manufacturing 40,000 vehicles per year by 2004 and 100,000 vehicles per year by 2006.

If cost targets can be met, fuel-cell-powered vehicles could command a significant share of the new vehicle market after 2010. Buses are likely to offer an important early market for fuel cells in transport. A number of fuel-cell buses have already been delivered to transit authorities for user evaluation. The first commercial products are expected by 2001.

Development work is under way on gasoline fuel processing for PEMFC systems, which would avoid the need for a new fuel infrastructure. But this approach represents a major technical challenge. Other options being pursued include direct methanol fuel cells, which preclude the need for a reformer, and the use of ethanol as a hydrogen source in place of methanol or gasoline. Toyota has claimed a substantial improvement in hydrogen-storage technology.

Emissions Reduction Potential: Fuel cells in conjunction with a gasoline reformer could be about 70 percent more efficient than current gasoline engines (but only slightly more efficient than a diesel hybrid drive train). A fuel-cell-powered vehicle would have about 45 percent lower CO_2 emissions than a conventional engine using the same fuel. Fuel-cell efficiency is highest for hydrogen and lowest for gasoline.

Barriers to Wider Use: Despite progress, fuel cells must overcome major hurdles before they can succeed commercially in the light-duty market. Costs must be reduced sharply, by as much as one to two orders of magnitude, for wide application in automobiles. Low-cost membranes and reductions in the size and cost of hydrogen reformers and on-board hydrogen storage equipment are particularly needed.

The lack of an existing fuel infrastructure is a barrier to the introduction of fuel-cell vehicles fuelled by hydrogen, methanol or natural gas, although methanol could take advantage of some of the existing infrastructure. Gasoline could do so as well, although the specifications for gasoline used in fuel cells are likely to differ from those for gasoline used in internal combustion engines.

Lack of demand for these technologies is also a significant barrier. It keeps the "technology learning" process from taking place, which would reduce costs (see Chapter 5).

Biofuels[28]

Description: Biofuels include biodiesel derived from vegetable oils and pyrolysis oils, bio-ethanol derived from cellulosic feedstocks[29], and bio-methanol derived from waste materials. Biodiesel and bio-ethanol can be used as blendstocks in conventional diesel fuel or gasoline and therefore can be used in conventional vehicles. Bio-ethanol and bio-methanol can also be used with butane to produce ethers such as ethyl-tertiary-butyl ether (ETBE) or methyl-tertiary-butyl ether (MTBE) that likewise can be blended into gasoline. With relatively small modifications, vehicles can operate on near-neat biodiesel, bio-ethanol or bio-methanol.

Under business-as-usual conditions, biofuels are not expected to contribute substantially to emissions reductions by 2020, although

---

28. *Sources for this section are AFIS 1999, the IEA Secretariat and contributions from IEA Member countries.*
29. *Cellulosic biomass is the major portion of plant materials such as wood, grass, organic wastes, and agricultural residues.*

they are near-term options in some regions. They are expected to play a much more significant role in the long term. But they fall into a special category of technology options whose near-term contribution could be made much larger if sufficient R&D is dedicated to reducing their cost in the near term.

Development Status: While the technical processes for these fuels are all well understood, cost remains high. Nonetheless, there is promise to achieve substantial cost reductions with significant levels of additional R&D and commercial-scale learning experience.

Emissions Reduction Potential: The emissions reduction potential is highly dependent on varying national circumstances. Agricultural countries are most able to take advantage of these technologies, especially if wood wastes and other agricultural surplus materials are available. Once these resources are consumed, further expansion of biofuels would depend on agricultural policies that would encourage the use of energy crops such as rapeseed, switchgrass and poplars. If an elastic supply of feedstocks is available, up to 3 percent of gasoline could be displaced with biofuels by 2010, and the potential after 2010 could be as high as 30 percent.

Barriers to Wider Use: High cost is the primary barrier to all biofuels. Technological advances are needed to reduce cost, especially for bioethanol from cellulosic feedstocks. Diesel vehicles may not be certified to operate on biodiesel fuel because of the potential for increased NO_x emissions. Another technical barrier is that methanol is very corrosive.

Assuming that further cost reductions are achieved, and other obstacles overcome, the primary barrier to widespread introduction of biofuels would be the large capital investments and investments in land needed to produce the fuels in great quantity. The current excess of global refining capacity and the expected low wholesale price of gasoline as a fuel competitor pose further barriers to such investments.

Some Promising Technologies for Clean Power Generation

Promising technologies for clean, and cleaner, power generation include fossil-fuel-based technologies, renewable-energy technologies, technologies to support extending the lives of nuclear plants, and fuel cells for stationary generation.

Natural-Gas-Fired Technology[30]

This category includes natural gas combined-cycle (NGCC) technology as well as the replacement or supplement of coal-fired generation with natural-gas-fired generation.

Description: NGCC power-generation systems are highly promising for emissions reduction in the near term. These systems are revolutionising the power industry. They have two important advantages over other options: high efficiency and low levels of pollutant and greenhouse gas emissions. Combined-cycle gas turbines (CCGTs) can achieve system efficiencies in the range of 50 percent or above.[31] This high efficiency is achieved by combining a gas turbine with a heat recovery steam generator.

The conversion of existing coal-fired power plants to operate on natural gas can significantly increase the efficiency of power generation and reduce carbon emissions. The simplest approach is *site repowering*, where the existing power plant site is reused with an entirely new NGCC system. This approach provides the highest cycle efficiency but requires a greater capital investment. In the more conventional approach of *steam turbine repowering*, a new gas turbine and heat recovery steam generator are used with the existing steam turbine and

---

30. *Sources for this section are contributions from the Working Party on Fossil Fuels, IEA 1994, EC 1997b, IEA 1995, IWG 1997 and PCAST 1997.*

31. *At present, power-plant efficiency is generally in the range of 32 to 35 percent on a higher heating value (HHV) basis. Some technologies do have higher efficiencies – for example, supercritical pulverised fuel combustion technology achieves efficiencies of 42 percent on a lower heating value (LHV) basis.*

auxiliary equipment. Because of equipment age and the fact that the steam turbine was designed for linkage with a coal-fired boiler, this approach results in lower efficiency than site repowering and hence a higher operating cost, but has a lower capital cost.

A gas turbine can also be coupled to the existing coal boiler, with 80 percent of the coal firing being maintained. Such an approach could reduce CO_2 emissions by 35 percent to 40 percent with only minor dislocation.

Development Status: Because of its highly competitive cost and its cleanliness and efficiency in conversion, and because the combustion turbine with or without combined-cycle technology is relatively inexpensive and can be put in place quickly, gas is the fuel of choice for new electricity capacity additions in the United States and Europe, enjoying a significant economic advantage over new coal plants. Efficiencies of NGCC systems are continuing to improve. Some NGCC plants are currently achieving efficiencies of 52 to 55 percent. Very efficient new systems, such as those recently built in the United Kingdom or Korea, are reaching conversion efficiencies approaching 60 percent.

Emissions Reduction Potential: Carbon dioxide emissions from NGCC plants are approximately 50 percent lower than emissions from conventional coal-fired plants. Switching from coal to gas in existing plants is limited primarily by natural gas availability and deliverability. Supply problems aside, if market and regulatory conditions favoured retirement of old coal plants and their replacement by NGCC plants, this option alone could meet the Kyoto targets within the electricity sectors of many countries, as the U.K. experience has indicated.

Barriers to Wider Use: Limitations on the availability and deliverability of natural gas restrict the addition of new gas-fired capacity in specific areas. In addition, economic penalties associated with mandated replacement of existing coal plants with natural gas plants can be high. Overcapacity in power generation at present limits the rate of market penetration of new NGCC plants.

The potential for social dislocation is a local, regional and national barrier to large-scale shifts from coal-fired generation to generation based on natural gas. Concern over the need for diversity in fuels and in fuel suppliers may also arise after 2010. Moreover, the price of gas could increase, which could shift the economics in favour of cleaner coal technologies.

Although NGCC technology is already widely used in many parts of the world, it still faces long-term technical challenges. For example, improved design and materials could further increase efficiency and allow higher operating temperatures.

More Efficient Coal Technologies[32]

Description: Coal-fired power plants will continue to be used in IEA Member countries for the foreseeable future, both because there are large, low-cost coal reserves in these countries and because natural gas is not a universally available resource. The key technologies for reducing greenhouse gas emissions from these plants in the near to medium term include advanced pulverised fuel (PF) technology and fluidised-bed combustion (FBC) technologies. Integrated coal-gasification combined-cycle (IGCC) technology will be important in the longer term.

Most of the world's coal-fired power-generation capacity depends on the use of PF boilers. In these systems, coal is ground into fine particles and injected with air into a combustor. Basic supercritical PF technologies can achieve efficiencies of about 42 percent on a lower heating value (LHV) basis. Advanced, "ultra-supercritical" boilers (for example, with steam conditions of 610°C and 25 MPa) can achieve efficiencies of 45 percent to 47 percent (LHV).

FBC technology uses crushed coal and limestone (the latter for sulphur removal), which are injected into a bed fluidised by the primary and

---

32. *Sources for this section are contributions from the Working Party on Fossil Fuels, EC 1997b, IEA 1997b, PCAST 1997, Scott and Nilsson 1999, and comments from IEA member countries.*

secondary combustion air. *Atmospheric* circulating designs can achieve 37 percent to 39 percent efficiency (LHV) in operation. All units currently operating are subcritical. Supercritical designs could help increase thermal efficiencies. *Pressurised* systems incorporate both gas and steam turbines in a combined cycle design. Using supercritical steam cycles leads to efficiencies of about 44 percent (LHV).

IGCC systems combine two established technologies: coal gasification for the production of synthesis gas (a gas mixture containing mainly carbon monoxide and hydrogen) and NGCC power production. Synthesis natural gas (syngas) obtained from the coal gasifier is used to drive gas turbines. The exhaust gas is used to generate steam that is converted to electricity by a steam-turbine bottoming cycle.

Development Status: Several coal combustion technologies are available today, and other advanced technologies are approaching commercial viability. Supercritical PF technology can now be considered "conventional". Ultra-supercritical technology is a recent innovation. Further advances in ultra-supercritical technology require the development of advanced materials. A coal-fired power station with an efficiency of 48 percent (LHV) is expected to be in operation by 2005, and an efficiency of more than 50 percent (LHV) should be possible by 2015. The development of "superalloys" based on nickel may allow efficiencies approaching 60 percent beyond 2015, though high cost may deter investment in such systems.

Atmospheric FBC technology (sub-critical and super-critical) is fully commercially available and has been applied in industrial heating, industrial and utility power generation, and CHP applications. Pressurised FBC technology is also commercially available, though operating experience is limited. Future developments are expected to increase efficiencies – perhaps to more than 50 percent (LHV).

Demonstration projects using IGCC technology are operating or under way worldwide, but the technology has not been widely deployed. Actual or expected efficiencies in five major demonstration projects (two in Europe and three in the United States) are on the order of 40 to 43 percent (LHV). Coal-fired IGCC technology is not currently competitive with large PF power stations, but development projects under way aim to increase efficiency and substantially reduce costs. An advanced process design could increase system efficiency to 47 to 49 percent by about 2005. In the long term, efficiencies greater than 60 percent may be possible. Commercialisation of more advanced designs (such as highly integrated designs) is unlikely until after 2010.

Emissions Reduction Potential: It is unlikely that advanced, high-efficiency coal technologies would achieve significant market penetration in IEA Member countries before 2010, largely because of competition from NGCC technology. By 2020, the introduction of high-efficiency coal-based technologies could reduce CO_2 emissions substantially, so long as these technologies did not displace moves to shift from coal to NGCC technologies (see Chapter 2).

Massive emissions reductions are possible today in developing countries and countries with economies in transition through the adoption of best practices in coal production and use. Even best practices in operations and maintenance could increase efficiency and reduce emissions from existing plants. In addition, in the absence of rapidly accelerating plant retirements in the OECD, countries such as China and India are likely to provide the greatest opportunities for major demonstration projects for clean coal technologies over the next few decades.

Barriers to Wider Use: Barriers to wider use of efficient coal-fired technologies are the slow growth in electricity demand, the widespread use of NGCC technology for new and replacement plants, and increasingly stringent emissions legislation. Gas-based power technologies are less expensive and they emit far less CO_2 per unit electricity than the best coal technologies. But the cost of coal is likely to remain low and the cost of gas may rise as demand for it increases. So, at some time in the future advanced coal technologies may be less expensive to use. Emissions could rise under these conditions unless CO_2 emissions can be captured and sequestered economically.

The emerging competitive market will put downward pressure on investment in both conventional and advanced coal technologies. Nevertheless, with continuing improvements in coal gasification and gas turbines, the added advantage of feedstock flexibility, and the opportunity to co-produce valuable byproducts from waste, advanced clean coal technologies could become economic choices for non-gas-based power production.

Advances in IGCC technology are still needed. Costs must be brought down and important technologies further developed. Techniques for filtering out highly corrosive impurities from the hot gas stream without unacceptable losses in efficiency are needed, as is better integration of the gasifier and combustion elements of the plant. Further demonstration is likely to be needed.

Renewable-Energy Technologies: Biomass and Wind

A number of renewable-energy technologies can be deployed in place of fossil-fuel-based generation. Their share of overall power generation is currently small, except for hydroelectric power plants, but it will expand. This section addresses two such technologies: biomass and wind. Solar technologies such as photovoltaics and concentrating solar power systems are not expected to contribute substantially to emissions reductions by 2020, although they are near-term options in certain regions. Solar energy is expected to play a much more significant role in the longer term.

Biomass[33]

Description: Biomass residues for power generation hold considerable potential to reduce emissions. Biomass fuels are derived from three sources: forestry, wood, and pulp-and-paper industry residues; agricultural and crop processing residues (such as sugar

33. *Sources for this section are EC 1997b, IWG 1997, Mintzer 1999, PCAST 1997, U.S. DOE 1997b, comments from the IEA Implementing Agreement on Bioenergy and comments from IEA Member countries.*

cane residues in Australia and elsewhere); and dedicated energy corps grown specifically for use as an energy feedstock. These fuels can be used to generate steam and electricity in thermal power plans, either alone or co-fired with coal. (In the longer term, biomass can also be converted into liquid or gaseous fuels.) There is little opportunity to expand biopower capacity with current technology where supplies of low-cost residues are limited, as they are in the United States. But prospects are good that biomass could play a major energy role through the use of advanced conversion and end-use technologies.

Co-firing biomass with coal represents an opportunity to reduce carbon emissions directly by substituting biomass-based renewable carbon for fossil carbon.

Development Status: For the near term, co-firing is the most cost-effective method of generating electric power from biomass resources. As an example, in the United States, this approach could replace at least 8 GW of coal-based generating capacity by 2010, and as much as 26 GW by 2020. Today, rapid expansion of biomass use is possible through the use of wood residues and dedicated feedstock supply systems. In the long term, prospects are good for new conversion technologies, such as biomass gasification.

Emissions Reduction Potential: Extensive demonstrations and trials have shown that biomass can replace up to 15 percent of the total input energy to a coal-fired plant with few modifications to existing equipment and hardware.

Barriers to Wider Use: Lack of feedstock availability locally is a barrier to specific projects.[34] High capital cost relative to fossil fuels – because of the lower energy density of fuel systems incorporating biomass – is a barrier to wider use of this technology. Lack of operational data for large-scale plant applications is

---

34. *In the near term, the potential biomass feedstocks available for a project can be defined as those within a radius of 50 to 100 kilometers.*

another barrier. Perceived technical risk and lack of well-established distribution systems for fuels are barriers in some markets. Moreover, potential conflicts between land use for bioenergy and competing demands to use the land for food production or other purposes could pose barriers to the wider use of biomass technologies in some countries.

Wind[35]

Description: Wind energy technology is one of the most promising renewable-energy technologies. Costs for electricity generated using wind have fallen dramatically in recent years. Wind energy technology is highly reliable and routinely achieves availabilities of 98 percent and higher.

The main use of wind energy is for electricity generation by large, grid-connected wind turbines in the range of 150 kW to 1.7 MW. Intermediate-size wind turbines are used in stand-alone systems, sometimes combined with other energy sources (such as photovoltaic cells, hydroelectric plants and diesel engines), with energy storage systems, or with both.

Development Status: The development of wind power systems has progressed rapidly since 1980. There were approximately 9,600 MW of wind capacity operating worldwide by the end of 1998; 2,100 MW of this were added in 1998. Most of the turbines installed in 1998 have rated capacities of 500 kW to 1 MW but systems with rated capacities of up to 2 MW are being deployed, primarily in Europe. The next generation of machines, expected to be in the size range of 2.5 to 3.5 MW, are likely to be installed offshore, at least in Europe.

Wind is now very close to competitive with alternatives in Europe, and it is less costly than alternatives in some niche markets.

---

35. *Sources for this section are contributions from the Implementing Agreement on Wind Turbine Systems, EC 1997b, IPCC 1996, IWG 1997, PCAST 1997, U.S. DOE 1997b, and comments from IEA Member countries.*

Despite dramatic advances, wind technology requires further R&D to reduce costs, improve performance and extend turbine lifetimes. For example, work is needed on lightweight adaptive structures, advanced materials for blades, and advanced controls and storage systems. Advances are also needed in reducing noise and in lowering the cost of foundations for offshore wind installations.

Emissions Reduction Potential: Wind is potentially a major player in emissions reduction. The wind resource is huge. Estimates indicate that, where wind is available, up to 10 to 20 percent of a region's electrical generation capacity could be supplied by wind without adverse economic or operational effects. Beyond this point, provisions would be needed for storage, backup and load management.

Barriers to Wider Use: Despite falling costs, wind energy is still more expensive than conventional energy sources outside of niche markets (and is expected to remain so in the near to mid-term), so cost remains the most serious barrier to its further development. The next largest barrier may be public concern over the visual impact of turbines. Noise generation is another problem and a barrier to development close to homes. Building turbines to operate at lower wind speeds (and thus on lower ground where the visual impact would be lessened) and to reduce noise would help.

Technologies for Nuclear Plant Optimisation and Life Extension[36]

Description: Life extension and optimisation of existing nuclear power plants are promising technologies for controlling the forecasted growth in emissions from power production. Nuclear power plants emit negligible amounts of greenhouse gases during operation. Their operating lives can be safely extended. In addition, nuclear and non-

---

36. *Sources for this section are PCAST 1997, U.S. DOE 1997b and comments from IEA Member countries.*

nuclear equipment in existing plants can be optimised to enable the plants to produce more electricity during their operating lifetimes.

Technical requirements for continued safe plant operation beyond the original design life include advanced digital instrumentation and control systems, extended fuel burnup systems, and efforts to optimise thermal and electrical efficiency. Major components, such as steam generators, may also need to be replaced. Plants may need to be upgraded to meet safety regulations. Advanced technologies for on-line monitoring of the condition of cables and conventional equipment (pumps, motors and so forth) may be needed to minimise production losses from unplanned outages. In-situ component and vessel annealing to repair cumulative radiation damage may also be needed.

Development Status: Under a joint programme between the U.S. Department of Energy and industry, a full-scale thermal annealing demonstration was conducted on a reactor pressure vessel in 1996. Advanced condition monitoring technologies have been developed for other applications and need to be adapted for application to nuclear power plants.

Emissions Reduction Potential: Maximising the lives of existing nuclear plants is potentially a cost-effective route to providing considerable amounts of carbon-free energy in the near to mid-term. For example, in the United States, optimising the operation of existing plants and extending their lives by 20 years could reduce carbon emissions by 60 to 110 million tons per year (compared with fossil-based generation of replacement electricity).

Barriers to Wider Use: Operating licenses for existing plants must be renewed in some countries, where the regulatory processes for this are new and uncertain. Research is needed on technologies to measure, diagnose and repair the effects of ageing on plant materials, components and systems. Sensors are needed that can operate reliably in harsh environments.

The capital investment required for plant upgrades may be large. Plants must still be economic to compete with other options for electricity generation, especially in liberalised markets.

Public opinion and political acceptability continue to be barriers to the operation of nuclear plants.

Fuel Cells for Stationary Generation[37,38]

Description: Fuel cells for stationary generation are promising technologies for reducing greenhouse gas emissions in the decade beyond 2010. Fuel cells have been developed for cogeneration (combined heat and power) applications and are being developed for residential and distributed power-generation applications. Hybrid fuel-cell/turbine systems are being developed for large distributed generation, industrial and utility applications.

Phosphoric acid fuel cells (PAFCs) represent the first generation of commercial fuel cells (200 kW). The development of this technology has enabled subsequent development of other fuel cells and of the needed infrastructure for fuel-cell technology.

The proton-exchange membrane fuel cell (PEMFC) is a low-temperature fuel cell with possible applications in transport, CHP systems, and distributed power generation. It offers the potential of low cost in mass production and power densities high enough for even demanding applications such as the automobile.

Developers of the molten carbonate fuel cell (MCFC) and the solid-oxide fuel cell (SOFC) seek 50 to 60 percent stand-alone efficiencies for distributed or centralised power-generation applications. These fuel cells may also be used with turbines in a combined-cycle arrangement. Both MCFCs and SOFCs have important potential in properly managed biomass systems.

---

37. *Sources for this section are contributions from the Implementing Agreement on Advanced Fuel Cells, EC 1997b, PCAST 1997 and comments from IEA Member countries.*
38. *Fuel cells for transport applications are addressed earlier in this chapter.*

Development Status: PAFCs are the only fuel-cell systems commercially available today. There are currently more than 200 PAFC systems installed worldwide, with a total generating capacity of about 50 MW. Other fuel cells are still in the development or demonstration phase, and there are still technical barriers to surmount. The first commercial PEMFC, SOFC and MCFC systems are expected within the next 4 to 6 years.

Low-temperature fuel cells have wide market opportunities in both transport and stationary CHP applications. Currently, the most promising low-temperature fuel cell is the PEMFC. Hydrogen is the natural fuel for low-temperature fuel cells. But PEMFCs might be launched in the market first with conventional hydrocarbon fuels and point-of-use fuel processors. For CHP applications, PEMFCs will be fuelled initially with natural gas that is reformed on site to a hydrogen-rich gaseous mixture that the fuel cell can use; fuel processors for such applications are commercially available.

Emissions Reduction Potential: The use of fuel cells for centralised electricity generation is unlikely within the next 10 years, although by 2015 Europe, Japan and the United States could all have significant installed capacity. There is potential for significant CO_2 savings because fuel cells are much more efficient than competing technologies in most applications. In power generation, for example, an advanced SOFC/gas turbine system is expected to operate at more than 70 percent electrical efficiency, producing only 50 to 70 percent of the CO_2 emitted from an equivalent CCGT plant.

Barriers to Wider Use: Cost is likely to be the major barrier to wide use of fuel cells for stationary generation. Production costs, even for the fuel cells approaching market readiness, are high, making them less competitive with established technologies. For example, costs for the installed PAFC systems are about 3,000 US$ per kW. (The manufacturer expects this cost to fall to about 1,200 US$ per kW once 3,000 units have been sold.) There are also technical barriers associated with the individual fuel cells. For example, the sensitivity of fuel-cell performance to impurities in the fuel stream is an important research topic for applications with coal, biomass or waste as their primary fuel. In addition, lack of awareness of the technology on the part of potential users is a barrier to wider use.

Some Promising Crosscutting Technologies

Combined Heat and Power[39]

Description: CHP is one of the most promising technologies for the near-term reduction of greenhouse gas emissions. It involves the joint production of heat (steam) and electricity. Both the heat and the electricity can be used on site, or surplus electricity can be sold back to the grid and surplus heat can be used in district heating (DH) and community energy systems.

There are substantial thermodynamic advantages to the joint production of heat and power that could greatly reduce generation losses from traditional power production and would reduce carbon emissions system-wide. Little additional fuel is required for electricity generation over that required for simple steam production, so overall efficiency is higher than with separate electricity generation and steam production. The most attractive use of CHP is where existing byproducts and waste can be burned (for example, wood chips, paper mill wastes, refinery gas), substituting for purchased fuel.

Development Status: Highly efficient lower-temperature CHP is well-developed technology for the commercial sector, but high-temperature CHP is still in its infancy. Only about 50 MW of high-temperature CHP are deployed in the European Union. Industrial processes where direct heat is needed in the range of 200°C to 800°C, such as those in bakeries, ceramics manufacturing, brick making and dairies, can make

---

39. *Sources for this section are contributions from the Implementing Agreement on District Heating and Cooling, EC 1997a, EC 1997b, IEA 1997a, IWG 1997, comments from IEA Member countries, and information published by Euroheat & Power and the European Marketing Group, District Heating and Cooling (http://www.eu-dhc.org).*

use of high-temperature CHP, although using it on a retrofit basis would require major plant changes such as replacement of existing furnaces. Currently, industrial CHP is used particularly widely in countries with major energy-intensive industries and town centres with an extremely dense heating and cooling load, as in the United States and Japan. It is also used widely in Europe; the most technologically developed DH/CHP systems are to be found in the Nordic countries and Germany. In a number of European countries, the CHP share of DH production has increased markedly over the past 20 years, reaching 79 percent in Sweden, 75 percent in Germany and 73 percent in Denmark.

Recent advances in the efficiency and cost-effectiveness of electricity generating technologies have allowed for the development of new CHP configurations that reduce size yet increase output.

Emissions Reduction Potential: In a CHP or DH/CHP system, the overall conversion efficiency of fuel energy to useful heat or power can be as high as 85 to 92 percent. It thus has the potential to reduce CO_2 emissions significantly compared with producing the heat and power separately from the same fuel. The American Council for an Energy Efficient Economy names CHP as one of the five most important energy-efficiency strategies for meeting the U.S. Kyoto commitments. In its strategy to promote CHP, the European Commission describes CHP as one of the few technologies that can offer a significant short- or medium-term contribution to energy efficiency, and thus CO_2 emissions stabilisation, in the European Union. The strategy estimates that the maximum electricity production potential of CHP in the European Union may be four times the amount generated from CHP in 1994. Full exploitation of this potential, replacing existing electricity and heat production plants, could reduce CO_2 emissions in the European Union by 9 percent compared with what they otherwise would have been. A U.K. study suggests that half of the CO_2 savings required through 2010 in the United Kingdom could be met cost-effectively with CHP.

Barriers to Wider Use: One of the largest barriers to CHP use is the difficulty of matching heat and electricity loads. In addition, the high initial costs of these systems could deter investment in the power-generation sector under deregulated markets as well as in industry. Operational problems, such as the effect of direct heating on product quality, could also deter industrial investment. Lack of experience with CHP in a given industrial sector could do the same. Depending on the locality, four additional barriers may be environmental permitting, which is often complex, time-consuming and uncertain; regulations that do not recognise CHP's overall energy efficiency or credit the emissions avoided from displaced electricity generation; discriminatory backup rates and interconnection fees charged by utilities; and unfavourable tax treatment and depreciation requirements.

Advanced Gas Turbines[40]

Description: Advanced turbine systems (ATSs) are a promising, crosscutting technology for the near and longer terms. The turbines are high-efficiency, next-generation, gas-fired turbines that will produce less carbon per kWh than technologies used in conventional power markets. They are being developed in two size classes: industrial gas turbines (approximately 5 MW and 15 MW) and turbines for utility combined-cycle systems (approximately 400 MW). These ATSs are one of the major low-carbon technologies for the industrial sector between now and 2010 because of their high efficiencies (greater than 40 percent) and their capability to cogenerate electricity and steam. These turbines are able to run on a variety of fuels and can be adapted for biomass and landfill gas fuels.

Development Status: Advanced turbine systems are poised to enter the market in 2000-2001. Some ATS manufacturers already have significant orders for their engines. Currently, most manufacturers are engaged in component and full system demonstration activities. Additional technology under development includes ceramic materials and coatings, low emission (less than 5 ppm NO_x) technology, and alloys. Some of these technologies will be commercially available in the ATS and spin-off engines by 2003.

---

40. *Sources for this section are IWG 1997 and U.S. DOE 1997b.*

Emissions Reduction Potential: When introduced in 2000-2001, ATSs will have CO_2 emissions 21 percent to 61 percent lower than conventional turbines. Efficiencies of industrial ATSs will exceed 40 percent and reach 60 percent for utility combined-cycle systems.

Barriers to Wider Use: With restructuring of electricity markets, many customers may delay decisions on investment in power generation and cogeneration until the regulatory situation is more settled. The low availability of capital in several end-user industries discourages investment in new technology. In addition, slow rates of capital stock turnover in industry will affect the rate of adoption of this technology.

Sensors and Controls[41]

Description: Sensors and controls are a promising crosscutting technology for energy end-use and power generation applications. Sensors and controls do not themselves save energy, but they increase the efficiencies of equipment and processes, thus reducing energy use and related emissions. Sensors and controls have applications in industry, where precise industrial process control is often limited by the lack of advanced sensor technology. They are also used in fossil energy extraction systems, building energy systems, advanced vehicle engines and other systems.

Development Status: Industrial process control and energy management are continuously evolving with new and improved sensors and controls and advances in information processing technologies. Use of advanced sensing and signal processing capabilities allows a progressive transition from localised control of a production process to full factory floor automation. The speed of this transition has been sharply accelerated over the past 5 years, aided by significant advances in fibreoptic, semiconductor, microfabrication and microprocessor technologies. With new devices becoming more reliable and cost-

---

41. Sources for this section are EC 1997b, U.S. DOE 1997b, and comments from IEA Member countries.

effective, the use of "smart" sensors in factory automation will become more widespread.

Emissions Reduction Potential: The use of sensors for industrial process control has already improved productivity substantially and has reduced energy consumption and CO_2 emissions. It is estimated that energy management systems using sensors and controls could save about 5 to 10 percent of process energy use in industry. Combustion control through closed-loop feedback has improved passenger-vehicle efficiency by an estimated 15 percent and fossil-fuel-burner efficiency by about 3 percent, with commensurate reductions in CO_2 emissions.

Barriers to Wider Use: The primary barrier to wide use of sensors and controls is technical: the lack of low-cost, robust and reliable sensors that are resistant to corrosive conditions and able to withstand high temperatures. The integration of sensors into control systems is also needed. High capital costs for new systems using sensors and controls is another barrier to wide use, as is the lack of readily available and accessible information on their potential economic and environmental benefits.

Power Electronics[42]

Description: Power electronics serve to upgrade power from distributed and intermittent sources to grid quality and to iron out disturbances to the grid that could result from end-use electro-technologies such as variable speed drives. Power electronics are thus important to renewables-based electricity generation, distributed generation and end-use electro-technologies. They themselves do not reduce CO_2 emissions but permit technologies to be used that can do so. For example, motors achieve variable speed capability, which results in increased efficiency, via power electronics. Power electronics are also used in electric vehicles. "Inverter" circuitry converts power generated using a number of alternative energy technologies – such as

---

42. *Sources for this section are IEA 1997b, U.S. DOE 1997b, IWG 1997 and comments from IEA Member countries.*

photovoltaics, wind energy systems and fuel cells – into alternating-current (AC) power.

Development Status: There are many commercial technologies, though improvements are still needed. Inverter technologies have recently been developed with improved efficiency, reliability and performance and reduced size and cost. A multi-level inverter has been developed that will allow 26 percent more energy to be extracted from photovoltaic or other renewable-energy sources.

Emissions Reduction Potential: The use of improved power electronics leads to reduced carbon emissions in virtually all electricity generating technologies and energy end-use technologies and can improve electricity grid operation and management.

Barriers to Wider Use: Cost and technical barriers stand in the way of wider use of power electronics. Smaller, lighter, more efficient, lower-cost inverters are required, and reliability, cost and electromagnetic compatibility must be improved.

Some Promising Technologies for Carbon Sequestration[43]

There appear to be no serious technical barriers to CO_2 sequestration, although high costs for CO_2 capture and uncertainties about environmental impacts and the long-term integrity of storage schemes remain as issues to be resolved. Technologies for CO_2 capture and sequestration are being demonstrated today. For example, Statoil, the Norwegian oil and gas company, is using state-of-the-art technology to capture CO_2 from the production of natural gas and sequester it in saline aquifers under the North Sea.

---

43. *Sources for this section are contributions from the Working Party on Fossil Fuels, PCAST 1997, U.S. DOE 1997b, comments from the IEA Implementing Agreement for a Greenhouse Gases R&D Programme, and comments from IEA Member countries.*

The primary sequestration technologies for emissions reduction through 2020 are CO_2 separation technologies and geologic storage of CO_2.

Carbon Dioxide Separation Technologies

Description: Efficient separation of CO_2 from flue gases is essential for any sequestration scheme. There are many techniques potentially usable for separation of CO_2 – for example, membranes and chemical and physical solvents. Cryogenics is another technique that can be used on streams with high concentrations of CO_2.

Development Status: Currently, various separation techniques using membranes are being tested in different parts of the world, and ongoing R&D efforts are aimed at developing more cost-efficient separation methods. Chemical and physical solvents are already used commercially for this purpose.

Emissions Reduction Potential: Carbon dioxide capture is applicable to large flue gas streams in energy-intensive industries and in power generation. It is able to deliver deep reductions in emissions while enabling continued use of fossil fuels. It is primarily applicable to new plants, as replacing existing coal plants with renewable-based or more efficient fossil-based technology would be cheaper than retrofitting existing the plants with CO_2 capture systems. In the case of coal plants, the obvious route to incorporating CO_2 capture into plant design is through coal gasification, which can be adapted to produce a pure CO_2 stream.

Barriers to Wider Use: The cost of separating CO_2 is a major obstacle to the wide use of available methods for CO_2 capture, as it largely exceeds the cost of transporting it, even over long distances. The need for appropriate storage sites could constrain the emissions reduction potential over the longer term.

Geologic Storage of Carbon Dioxide

Description: Geologic storage of CO_2 is the most promising sequestration technology for the near term. It involves capturing the gas and injecting it into subsurface repositories such as deep coal beds; depleted oil and gas reservoirs; and deep, confined saline aquifers. Other options aside from geologic storage are also being investigated, including deep ocean storage and use of fertilisation to enhance the ocean carbon sink.

Development Status. The technology for subsurface injection is readily adaptable from the petroleum industry. That industry uses technologies for drilling and completion of injection wells, compression and long-distance transport of gases, and characterisation of subsurface reservoirs. It has experience with CO_2 injection for enhanced oil recovery. Natural gas, as an analogous model, is routinely transported and stored in subsurface reservoirs and aquifers.

Amine absorption, in combination with CO_2 storage in saline aquifers, has been used since 1996 in the Sleipner Vest project in the Sleipner Gas field, located in the Norwegian part of the North Sea. In this case, CO_2 is separated directly from the well stream before the gas is further processed and exported. About 1 million metric tons of CO_2 are separated and injected into the reservoir annually.

Injection of CO_2 into partially depleted oil reservoirs is being used for enhanced oil recovery at about 70 sites worldwide. Injection into deep, unmineable coal beds to recover coalbed methane is under active investigation in a number of countries.

Emissions Reduction Potential: Long-term storage in geological repositories will reduce greenhouse gas emissions by sequestering them from the atmosphere. Injection into depleted oil and gas reservoirs and deep coal beds could store CO_2 and also yield commercially valuable hydrocarbons. In a study carried out for the European Commission, underground aquifer storage capacity in the North Sea alone was estimated to be adequate to store up to 200 to 250 years of CO_2 emissions from OECD-Europe, at 1990 emission rates.

Barriers to Wider Use: The main practical barrier to wider use of geologic storage is the cost of separating CO_2 from dilute flue gas streams.

CHAPTER 4
TECHNOLOGIES AND R&D
FOR THE LONG TERM: SOME
PROMISING DIRECTIONS

Introduction

As is clear from the previous chapter, there are many technologies that can reduce greenhouse gas emissions from energy production and use within the Kyoto time frame and shortly beyond (to 2020). But these technologies are only part of the story. The emissions reductions that can be achieved in this time frame will not be sufficient to adequately slow the increasing concentration of greenhouse gases in the atmosphere. As important as the Kyoto Protocol will be as a first step, additional commitments will be needed beyond 2010 to achieve the further emissions reductions likely to be required.

Governments should therefore not focus *solely* on the short term. A vigorous effort to realise short-term gains will help build momentum for a sustained effort to reduce emissions over the long term. But this short-term effort must be complemented by a strong push to develop and deploy technologies that will have an effect only beyond the Kyoto time frame.

This longer-term push calls for action in two broad categories: (1) long-term R&D, and (2) the facilitation of deployment once advanced technologies are developed. This chapter, which draws heavily on U.S. DOE 1997b, describes some of the technologies and technological advances that could be available after 2020 as a result of long-term support for R&D. It also outlines some of the needed research. Support for deployment – for near-term technologies as well as those that will take longer to make their way into the market – is treated in Chapter 5. Chapter 6 addresses governments' role in long-term R&D.

As was the case with Chapter 3, the technologies mentioned in this chapter are not intended as a complete list of possibilities. Many more advances in technologies and energy systems could be described. Fuller discussions are available in the sources used for this chapter. Rather, this chapter provides a flavour of some of the major advances that could occur with a sustained commitment to R&D, particularly long-term R&D.

Long-Term R&D: "Post-Kyoto" Technologies that Require Development

Many of the most promising technologies to reduce carbon emissions substantially require considerable applied R&D before they will be commercially feasible. There are still others that are only at the conceptual stage but can be developed with further research. The significance of these long-term efforts lies in important technologies that were largely left out of the previous chapter. These include continued advances in end-use efficiency, advanced renewable-energy technologies, hydrogen technologies, nuclear fusion, and continued advances in clean coal technology and carbon sequestration. Technologies such as these, if they can be made economic, will provide the opportunity to achieve large reductions in emissions from energy production and use in the long term – reductions not possible in the Kyoto time frame – while allowing for continued economic growth and growth in the use of energy services.

Some of the technologies described here can be easily incorporated into the existing energy system. Others will require infrastructure changes ranging from minor to extensive. Still others will trigger major shifts in how energy is produced and used.

Energy Efficiency – Buildings[44]

Chapter 3 stressed the need to increase the use of efficient, commercial and near-commercial versions of building equipment and of the more easily modified components and features (such as windows and insulation) of building structures. Further applied R&D and advances in the underlying basic sciences, such as materials science, could result in advanced, "best practice" buildings incorporating features such as the following:

- computer-based building design and optimisation;
- manufactured wall systems with integrated superinsulation and electrochromic windows;
- "superwindows" optimised for orientation (to take advantage of natural lighting), external temperature and internal needs;
- integrated natural and electric lighting systems (for commercial buildings, highly-efficient centralised electric light sources combined with tracking daylight collectors connected to "piped" light-distribution systems);
- photovoltaic roof shingles, reflective roofing and strategic positioning of trees to reduce cooling costs;
- fuel cells for power generation and space conditioning;
- sensor-controlled ventilation systems with air filtration and heat exchange;
- advanced building control systems incorporating "smart" technology to closely match energy and water supply for efficient, multi-functional and integrated appliances and to match ambient conditions with need;
- energy storage systems;
- advanced district heating and cooling systems.

---

44. *Sources for this section are contributions from the Implementing Agreement on Energy Conservation in Buildings and Community Systems, IWG 1997 and PCAST 1997.*

Future buildings can be completely self-powered through the use of fuel cells, small turbines, photovoltaic building components (panels, shingles and so forth) and energy storage systems. Excess electricity can be generated for sale to the grid or for use in charging electric vehicles.

A future "sustainable" building can be envisioned that would have a minimal impact on the indoor, local, regional and global environment. It would use recyclable construction materials, would consume a minimum of non-renewable heat and electricity, and would employ heat recovery and heat cascading. It would be connected to wastewater cleaning/recycling and to waste management/recycling, would have high energy efficiency, and would provide good thermal comfort, indoor air quality and lighting throughout the year.

In the long term, if greenhouse gas emissions attributable to the building sector are to be significantly reduced, incorporation of features such as those described above must be the norm, not the exception. To make this happen, work is needed to integrate energy-efficiency practices and principles into broader building-sector policies (codes, standards and so forth) as described in Chapter 5. Because the lifetimes of buildings are so long, actions to influence building design, construction and refurbishment are high-leverage activities for the long term. In addition, long-term R&D is required to provide the advanced appliances, equipment, components, systems and design tools needed to create advanced buildings. Examples of areas in which such R&D is needed are on-site power generation, system optimisation, advanced sensors and smart controls, energy design and diagnostic tools, automated diagnostics, superinsulations, adaptive building materials and envelope systems, innovative thermal distribution networks, recycled materials, integrated multifunction appliances, innovative lighting and new materials for appliances.

Energy Efficiency – Industry[45]

As Chapter 3 noted, crosscutting industrial technologies can reduce industrial energy use in existing and new facilities in the near term. In the longer term, the opportunity to influence new plant design, and R&D to develop advanced technologies for both specific processes and crosscutting needs, can result in even greater reductions. Novel concepts such as integration of industrial facilities with other plants and with facilities for power supply and waste management could lead to "zero-emission" systems.

The efficiency of energy conversion processes in industry can be improved by incorporating the best available technologies – those described in Chapter 3 as well as others – in a systems approach, particularly for new plants. In the longer term, fuel cells and gasification of biomass and in-plant residues (such as black liquor in the forest products industry) are likely to have a large impact.

In addition to the energy conversion improvements mentioned above, developing new, more efficient processes can also substantially reduce emissions from energy use in industrial processes. Such processes can encourage new, higher-quality products while generating less waste and fewer undesirable byproducts. Opportunities exist to improve process efficiency via advances such as more selective catalysts, further developments in advanced separations, improved materials and improved electric motor systems. A particularly attractive longer-term opportunity is the use of biotechnology and bio-derived chemicals and materials.

Increased fundamental understanding in enabling sciences such as chemistry, metallurgy and biotechnology will allow the development of novel manufacturing processes. This knowledge, along with enabling technologies such as advanced modelling and simulation, improved industrial materials, and advanced sensors and intelligent control systems, can result in major incremental improvements and lead to

---

45. *Sources for this section are IWG 1997 and U.S. DOE 1997b.*

fundamental breakthroughs. Likewise, developing and demonstrating micromanufacturing systems (such as mini-mills and micro-chemical reactors) for flexible process configuration and on-site/just-in-place (similar to just-in-time) manufacturing can reduce emissions in the long term. Decentralised manufacturing using locally distributed resources offers the advantage of reduced transport of raw materials and finished goods.

Resource recovery and utilisation offer further savings. An advanced concept is an industrial ecology, in which a community of producers and consumers performs in a closed system. Fossil energy is conserved or energy is obtained from sources that do not give rise to greenhouse gas emissions; materials are reused or recycled. Through technological advances, the raw materials and resources needed for manufacturing can be obtained by designing products for ease of disassembly and reuse, using more recycled materials in finished goods, and selecting raw materials to eliminate waste discharge or undesirable byproducts. Examples of developments that could facilitate this approach are new polymers, composites, and fibres and advanced ceramics engineering techniques. Another approach is to substitute materials such as biomass feedstocks for petroleum feedstocks in producing chemicals. Some longer-term technological approaches could use CO_2 as a feedstock and reductants that do not lead to greenhouse gas emissions as substitutes for carbon. These approaches represent fundamental changes in the way raw materials are obtained, the properties they exhibit and the way they are used in the design process.

Energy Efficiency – Transport[46]

The transport sector has many technologies ready, or almost ready, to improve fuel efficiency and reduce emissions, but these technologies have not been fully applied because of a lack of demand or their efficiency gains have been negated by larger vehicle size and higher vehicle power. Thus, with continuing R&D, the technologies described

---

46. *Sources for this section are IWG 1997, U.S. DOE 1997b and the IEA Secretariat.*

in Chapter 3 – such as lean NO_x catalysts for diesel engines and lightweight materials – are expected to be available in the relatively near term. One exception to this statement is advanced fuel-cell vehicles. Continuing intensive R&D will be needed to overcome technical barriers and reduce costs. Biofuels represent the case of a technology that is available but is not expected to contribute substantially to reducing greenhouse gas emissions in the near term because of high cost. A significantly increased R&D effort that brings down biofuel costs quickly could greatly increase the near-term contribution of this technology.

In the long term, new technologies hold out the promise of spectacular reductions in greenhouse gas emissions from transport. The combination of advanced fuel-cell technology and an infrastructure for supplying hydrogen offers the potential for a pollution-free propulsion system, depending on how the hydrogen is produced. The lack of adequate infrastructure and of on-board storage technologies for hydrogen are the greatest obstacles to its use as a transport fuel.

Use of fuel cells in heavy trucks and locomotives will require a breakthrough in hydrogen production, distribution or on-board storage, or a breakthrough in reforming technology, before it will be competitive with the diesel engine. The drive-cycle thermal efficiency of current heavy-duty diesel-truck drivetrains is roughly the same (35 to 40 percent) as that for current methanol steam-reforming fuel-cell drivetrains (including the electric motor/controller and battery). Therefore, there is likely to be little incentive to switch heavy trucks to fuel cells until hydrogen fuel cells, with higher efficiencies (45 to 50 percent), become competitive and breakthroughs in hydrogen production, distribution and storage occur. Fuel cells may find acceptance in locomotives before they do in trucks.

For heavy-duty diesel engines, there is a strong inverse relationship between efficiency and reduction of non-CO_2 emissions. Many engine design options currently available to manufacturers for emissions reduction involve a fuel economy penalty of 10 to 20 percent. Significant technology advances are needed to allow the trend toward higher diesel-engine efficiency to continue in the face of increasing concern over non-CO_2 diesel engine emissions.

Clean Power Generation – Fossil Fuels[47]

Fossil energy dominates the world's energy supplies and is likely to do so for the foreseeable future. It provides about 90 percent of the world's energy, with oil providing 41 percent, natural gas 22 percent and coal 27 percent (IEA 1998b). Because OECD economies are so heavily reliant on each of these three fuels, it would take at least two decades of dedicated technical and infrastructure development to alter this fuel mix substantially.

Fossil Resource Development

Separation and conversion processes used in crude oil refining are energy intensive, often relying on combustion of light hydrocarbons that produces CO_2. Fugitive emissions of light hydrocarbon gases from refinery operations are another source of greenhouse gas emissions. Promising solutions to reduce these emissions include improved catalysts, advanced separation methods (membranes), mild pretreatment processes, refinery process optimisation and advanced sensors.

Natural gas has a fundamental advantage over oil and coal in terms of reducing carbon emissions because of its lower carbon-to-hydrogen ratio. Economical conversion of natural gas streams to liquid products such as fuels and commodity chemicals will allow full use of natural gas supplies while addressing the issue of greenhouse gas emissions. Recent studies have indicated that diesel fuels produced from natural gas are significantly less polluting than petroleum-derived diesel. Research and technology advances are needed to improve conversion processes for liquefied natural gas. For example, better tools are needed for modelling of fluid dynamics and non-equilibrium chemistry. Better sensors for automated process control and optimisation are needed, as are advanced selective catalytic materials.

---

47. *The source for this section is U.S. DOE 1997b.*

Increased natural gas production in the short term will result from incremental improvements to current practices. In the intermediate time frame, conversion of natural gas from remote sites into a high-quality liquid fuel that is moved by pipeline to a distribution infrastructure would have the greatest impact on increasing natural gas production.

In the longer term, entirely new technology will be required to map and produce undersea methane hydrates (physical combinations of methane and water) and perhaps even deep-source gas. Methane hydrates represent a potential energy source greater than all known oil, gas and coal deposits combined. Research challenges include developing innovative drilling and subsurface diagnostics, improving diagnostics to detect gas hydrates and deep gas resources, and developing technology to recover gas from hydrates and deep gas sources (no technology currently exists to extract methane from undersea hydrate deposits).

High-Efficiency Coal-Based Technologies

In the long term, highly efficient coal-based technologies could significantly reduce carbon emissions. Primary technology development is needed – in support of advances in the technologies described in Chapter 3 – in areas such as high-temperature materials that are stable and that resist corrosion, erosion and decrepitation so that they can be used in heat exchangers, turbine components, particulate filters and SO_x removal equipment. Other challenges include the use of alternative working fluids for turbine and heat-exchange cycles, cycle optimisation, environmental control technologies with low energy penalties, and solids handling.

As described in Chapter 3, integrated gasification combined-cycle (IGCC) technology could achieve an efficiency of 60 percent in the long term. It can be used to produce a variety of commodity and premium products as well as electricity. In "co-production" mode, feedstocks can be processed before gasification to extract valuable components, or the synthesis gas can be converted to products. Valuable precursors from feedstocks such as coal can be extracted to manufacture high-strength, lightweight carbon fibres and anode coke. Clean synthesis gas can be catalytically converted into environmentally superior transport fuels, high-value chemicals or hydrogen.

R&D challenges for further development of advanced IGCC and co-production technology include reducing the cost of synthesis gas for both power and fuels production; developing catalysts and sorbents that are mechanically strong and have high activity; optimising the design of process facilities; improving materials for high-temperature gas turbines; and developing innovative, cost-effective methods for using or sequestering CO_2.

Low-Carbon Fuels

When used in advanced fossil generating technologies, low- or no-carbon fuels – such as natural gas, synthesis gas and hydrogen – could lead to a significant reduction in carbon emissions after 2020. Fuel cells and gas turbines are currently in use, taking advantage of plentiful supplies of natural gas. But neither technology has reached its potential. For gas turbines, lowering costs and developing advanced turbine systems and gasification adaptation technologies will require improvements in blade cooling and materials. Considerable development is required before advanced technologies will be capable of high performance when firing hydrogen. In addition, further improvements are needed to avoid increasing NO_x emissions as operating temperatures rise.

As described in Chapter 3, fuel cells represent another potential gas-fuelled technology. Recent successes have shown the potential to build larger stacks of fuel cells, leading to larger generating capacity, but costs need to be reduced to make them competitive. Performance targets include developing market-ready fuel- cell systems with efficiencies higher than 50 percent, adapting them to operate on synthesis gases from coal and other solids, and validating the hybrid fuel cell/advanced gas turbine system that could have efficiencies approaching 70 percent by 2010. Demonstration of advanced turbines on hydrogen alone or in some hybrid cycle is expected to occur between 2020 and 2030.

Energyplexes

Ultimately, ultra high-efficiency, zero-carbon-emissions systems can be envisioned that would take advantage of synergies between energy generation, fuels production and chemicals production by integrating these processes into a single entity, an "energyplex". An energyplex would incorporate a series of modular plants capable of co-producing power and chemicals or fuels that can be integrated to use local sources of carbon (coal, biomass, municipal solid waste) as fuel and feedstocks. With the incorporation of modules for capture and sequestration of CO_2, energyplexes would have essentially no carbon emissions.

These complexes would optimise the entire cycle of carbon use by incorporating co-processing concepts, the integral capture of CO_2 and the incorporation of carbon into useful products or carbon sequestration. This is a long-term, futuristic concept that challenges the R&D community to make significant technology breakthroughs in areas such as novel industrial process configurations, novel power cycles and co-production of heat and power, with suitable energy-efficient reuse or storage options for carbon and CO_2. Representative technologies include IGCC systems, coal liquefaction, fuel cell/gas turbine bottoming cycle combinations with efficiencies of more than 70 percent and integral capture of CO_2, power systems with alternative working fluids, high-temperature oxygen separation membranes, and advanced oxygen production techniques.

Clean Power Generation – Renewables[48]

It is quite likely that renewable-energy technologies will supply an increasing share of global primary energy during the coming decades. Some analysts have suggested that the use of these technologies will

---

48. *Sources for this section are IWG 1997; Mintzer 1999; U.S. DOE 1997b; and contributions from the Implementing Agreements on Geothermal Energy Technology, Solar Photovoltaics, Solar Power and Chemical Energy Systems, and Wind Power Technologies.*

expand to become the world's principal source of primary energy in the latter half of the twenty-first century.

The success of current and planned R&D efforts will be among the key factors that determine whether advanced renewable-energy technologies capture a significantly higher share of global primary energy supply in the longer term. The principal goals for these R&D efforts include reducing the cost of energy production from renewable resources, increasing the quality of energy delivered and the reliability of renewable-energy supplies, and improving the matching of energy supply with end-user demand so as to reduce costs and losses in energy transport and distribution. Energy storage technologies are also important to greater use of renewable-energy technologies.

For some technologies – solar photovoltaics in particular – significant cost reductions are expected from development of new materials, improved integration of balance-of-system components into specific applications, and development of large-scale manufacturing and production facilities.

The following are some examples of R&D efforts that hold promise for the long term:

- In the long term, **biomass use for electricity supply** may depend on the development of advanced power cycles using gasification technology. Biomass-integrated gasification/gas turbine (BIG/GT) systems are the focus of substantial current attention in this regard. Many analysts believe that conversion efficiencies of 40 to 45 percent are achievable in the intermediate to longer term. A number of small BIG/GT demonstration projects are now under development in Sweden, Jamaica and Brazil, and new projects have been announced in Denmark, the United Kingdom, Italy, Belgium, the Netherlands and the United States. Additional development activity is focused on improvements in biogas technology, aimed at simple, efficient, inexpensive and reliable anaerobic digesters for use in rural areas of developing countries and at industrial biogasification plants that could produce biogas for combined heat and power applications or compressed gas for vehicles.

■ The most critical element of long-term R&D on **wind electric systems** is in the area of wind turbine design. The overall goal is to produce stable, reliable machines that use as little as half the material content of current turbines, at manufacturing costs approximately 75 percent of today's levels. Key R&D challenges include optimising wind turbine systems for operation over 30 years in fatigue-driven environments with minimal or no component replacements, based on knowledge of wind flow, operative aerodynamics, structural dynamics, and optimal control of turbines and wind farms. As turbine size increases, understanding the interactions between the wind input and various components, and among components, is a fundamental challenge. Larger contributions from wind in the long term may require addressing the variable nature of the wind resource through measures such as modified systems operation, hybrids with other technologies and energy storage.

■ The most exciting research prospects for **photovoltaic power systems** include the development of multi-junction, stabilised, thin-film solar cells using amorphous silicon or tertiary semiconductors. In addition, substantial progress is expected in the area of advanced concentrator designs coupled to high-efficiency, single-crystal solar cells. There is considerable scope for long-term progress on advanced manufacturing processes and balance-of-plant components. Recent efforts to design photovoltaic systems into the "skin" of conventional building envelopes may also substantially reduce the net cost of these systems and increase their attractiveness to end users.

■ Long-term R&D programmes are focused on the production of solar fuels for a range of uses, employing **thermochemical conversion of solar energy** into chemical energy carriers (including hydrogen, synthesis gas and metals). For example, water-splitting thermochemical cycles based on reduction of metal oxides are under investigation, as is solar reforming of natural gas.

■ Research to demonstrate the technical feasibility of **advanced solar photoconversion** technologies is needed. These technologies use the energy of sunlight to produce fuels, materials, chemicals and electricity directly from renewable sources such as water, CO_2 and nitrogen. Most of these technologies – involving photobiological, photochemical and photoelectrochemical approaches – are still in the fundamental research stage. They offer the promise of natural and artificial photosynthesis processes that could, for example, produce hydrogen from water or biomass and produce biodiesel, methane and methanol from water, waste and CO_2.

■ The levelised costs of **concentrated solar power,** formerly known as solar thermal electric power, have come down by more than 50 percent over the past 15 years. Combining this technology with energy storage or with non-intermittent generation allows electricity production to continue when solar energy is not available. Extensive research is under way to improve the performance of parabolic trough and dish concentrator systems for conversion of sunlight into electricity at large, grid-connected, central-station facilities. If progress can be made through R&D to reduce the cost and improve the reliability of components, concentrated solar power systems can make a significant contribution to electric power generation.

■ Advanced **hydropower** technology will improve on available technologies for producing hydroelectric power by eliminating adverse environmental effects (such as fish entrainment/ impingement and the alteration of downstream water quality and quantity) and increasing operational efficiencies. R&D challenges include developing computational fluid dynamics models of forces inside hydropower turbines that can predict stress levels on fish, and demonstrating cost-effective retrofitting of existing plants using new technology.

■ Extraction of heat from **geothermal "hot dry rock"** requires pumping water into the rock at high pressure to widen existing fissures and create an underground reservoir. If the associated technology can be developed in a cost-competitive manner, it opens up the prospect of exploiting a very large, widely available resource capable of supplying both heat and power. Some analysts estimate that a commercial hot dry rock technology could supply up to 15 percent of the electricity needs of many countries.

Clean Power Generation – Nuclear[49]

Nuclear Fission

A new generation of fission reactors is required to replace or supplement existing light-water reactors (LWRs) after 2020. Evolutionary LWRs of standardised design are available. In the long term, advanced reactors and fuel cycles would aim to extract far more of the energy available in the uranium fuel than that extracted by current reactors, while operating at high levels of safety, reducing or eliminating the potential for proliferation of nuclear materials and limiting waste generation. Advanced reactor concepts, such as the liquid-metal-cooled, fast spectrum reactor, require large-scale technical demonstrations before commercialisation.

The concept of a nuclear-hydrogen cycle, in which nuclear heat is used to produce hydrogen, which is then used as an energy source, has considerable potential. R&D on fission heat-to-hydrogen conversion technologies and on hydrogen distribution technologies has the potential to produce fundamental changes in the world's approach to energy supply.

Thermonuclear Fusion

Fusion power systems hold substantial promise as an energy resource for the longer term. According to the IEA/CERT Fusion Power Co-ordinating Committee, fusion power can now be viewed as ready technically for embarking on its demonstration phase with the

---

49. *Sources for this section are contributions from the Fusion Power Co-ordinating Committee and U.S. DOE 1997b.*

International Thermonuclear Experimental Reactor (ITER). The technical time scale for achieving the successful operation of a demonstration power station could be about 20 years after the start of operation of ITER. Budgetary and procedural delays could stretch this time scale out to as much as 40 years.

Crosscutting Technologies[50]

Hydrogen Technologies

Hydrogen is a carbon-free energy carrier that has potential uses in many applications, as evidenced by its mention in several preceding sections of this chapter. For example, it can fuel vehicles, provide process heat for industrial processes, supply domestic heating needs through cogeneration or heat recovery systems, and fuel power plants for centralised or distributed generation. It burns cleanly and efficiently and can be used in modified conventional combustors to ease the transition to a completely new energy infrastructure based on using hydrogen in fuel cells for energy conversion.

If the greenhouse gas emissions benefits of relying on hydrogen as an energy carrier are to be fully realised, the hydrogen must be produced efficiently from a primary source (such as fossil fuels, municipal solid waste, biomass or water). The level of emissions reduction compared with conventional technologies will depend on how the hydrogen is produced. When it is produced via electrolysis of water using nuclear or renewable electricity, CO_2 is essentially absent from the fuel cycle.

Expanded R&D is needed on biological, thermochemical and electrochemical processes for producing hydrogen. Research is also needed on hydrogen storage technologies such as those based on innovative materials – for example, carbon fibres and structures and metal hydrides.

---

50. *Sources for this section are contributions from the Implementing Agreement on Assessing the Impacts of High-Temperature Superconductivity on the Electric Power Sector, U.S. DOE 1997b, and comments from IEA Member countries.*

Sensors and Controls

Sensors and controls play an important role in nearly every technology and system related to CO_2 emissions reduction and sequestration. A large variety of novel sensor technologies that are robust, sensitive, cost- effective and capable of supporting real-time control, along with commensurate advances in data analysis and fusion for control, will be required to maximise efficiency in operating advanced energy supply and end-use technologies as well as carbon capture and sequestration technologies. Examples of advanced technologies needed include physical sensors for in-situ measurement of process parameters; sensors for chemical speciation in hostile process and combustion control environments; sensors with embedded self-diagnostics and calibration; and enabling technologies for real-time data analysis such as pattern recognition, artificial intelligence and fuzzy logic. Research is also needed in the development of multi-analyte sensors and in the integration of sensors and microtechnologies, as in microflow devices. Further research is necessary to the development of "smart controllers" that couple a multitude of sensors or sensor arrays to sophisticated data analysis systems that can provide on-line process control and improve process efficiencies. Technical challenges to be overcome include materials development and packaging methods for harsh environments; signal processing to extract useful and repeatable information from low-level, noisy measurements; and methods of communication that permit whole systems of sensors and controls from multiple manufacturers to operate together.

Improved Transmission and Distribution Systems

Electrical transmission and distribution system improvements are essential to enable deployment of alternative electrical generating technologies, particularly for large-scale development of remote resources such as renewables. Power system component development to reduce losses from transmission and distribution systems also offers significant opportunities to reduce greenhouse gas emissions.

Advances are needed to better use existing systems. Advances are also needed in new systems and components, such as those required for distributed utility and generation concepts. R&D is needed on automated system control technologies and high-strength overhead line conductors to increase the capacity of existing systems. Developments in power electronics – including wide-band semiconductors for high-power switching devices and advanced converter designs – are needed to improve power management on existing systems and to enable high-voltage direct-current (DC) transmission for long-distance power transfers. Improvements in superconducting materials and associated refrigeration technologies will lead to the development of superconducting cables and power transformers that offer half the energy losses and many times the capacity of conventional devices while taking up less space and reducing environmental impact. Further development of superconducting generators, motors and transmission cables, and of low-cost methods for manufacturing amorphous metal materials for high-efficiency distribution transformers, is needed.

Energy Storage

Advanced energy storage technologies include mechanical processes (flywheels, pneumatic systems), electrochemical technologies (advanced batteries, reversible fuel cells, hydrogen) and purely electrical technologies (ultracapacitors, superconducting magnetic storage). Their greatest value for electricity systems is that they can enable better use of intermittent renewable-energy sources, such as solar energy and wind, that produce no direct CO_2 emissions.

The major challenge for all storage technologies is cost reduction. Key R&D needs include developing new electrocatalysts, new electrode materials and new structural materials for electrochemical systems; magnetic bearings, better fail-safe designs and lightweight containment, and composite rotors with higher specific-energy for flywheels; better corrosion-resistant materials for batteries with higher power density; commercial high-temperature superconductors (operating at liquid nitrogen temperatures) for superconducting magnetic energy systems; higher energy-density ultracapacitors for light-duty vehicles; and improved power conditioning systems.

Carbon Sequestration[51]

Research and development is needed on several advanced approaches to carbon sequestration:

- Augmented **ocean fertilisation**, which would enhance the capability of the ocean to sequester CO_2 by enriching nutrient-poor regions of the ocean floor with iron or nitrogen. This enrichment is a first step toward stimulating phytoplankton growth, enabling carbon deposition on the ocean floor and CO_2 drawdown.

- Advanced **chemical and biological sequestration**, which is aimed at permanent stable sequestration and recycling of carbon into new fuels and chemical feedstocks. Emissions are reduced through converting CO_2 into an environmentally benign product while generating liquid fuels, generating hydrogen fuel and converting CO_2 into organic compounds. Representative technologies include chemical sequestration as mineral carbonate, direct solar conversion of CO_2 to methanol, advanced conversion of coal to hydrogen and conversion of CO_2 into reusable biomass using microalgae.

- **Elemental carbon sequestration**, which is based on production of hydrogen from fossil and carbonaceous fuels (for example, by thermal decomposition of natural gas) followed by sequestration of the particulate carbon formed in the reaction. This process is an alternative to conventional steam reforming and subsequent sequestration of CO_2. Thermal decomposition of methane to produce carbon black is conventional technology. Production of

---

51. Sources for this section are contributions from the Working Party on Fossil Fuels, U.S. DOE 1997b, and comments from the IEA Implementing Agreement for a Greenhouse Gases R&D Programme.

hydrogen from methane decomposition has been demonstrated in a catalysed fluidised bed, and plasma decomposition has been performed in a pilot plant. But these are relatively inefficient processes and development of reactors that can produce hydrogen continuously at an efficiency of about 70 percent is needed. Research is also needed on using large amounts of carbon as a material commodity – for example, in construction applications.

■ **Carbon sequestration in soils**, which involves implementation of appropriate soil management practices that also increase agricultural productivity. These include minimum tillage agriculture, increased return of crop residues to the soil, use of irrigation and fertilisers at levels that maximise crop and root biomass, return of agricultural lands to forests and grasslands, and plant breeding and genetics to increase below-ground carbon storage. These techniques are available now, but research is needed to evaluate their effectiveness in sequestering carbon.

■ **Deep ocean sequestration** of CO_2 in gas hydrates, which could be a stable form of sequestration. The CO_2 sequestered in the deep ocean can also be stored in a liquid form or as a solution. R&D is needed in performance assessment, deep ocean science and engineering, fate and transport geochemistry, and three-dimensional characterisation and monitoring.

Agriculture and Forestry Efficiency[52]

Research is needed on several technologies and systems that can reduce greenhouse gas emissions from the agriculture and forestry sectors.

The conversion of biomass to materials and products can provide a modest but significant reduction in greenhouse gas emissions by lowering the use of petroleum-based feedstocks and reducing consumption of energy and fertiliser. Representative technologies are

---

52. *The source for this section is U.S. DOE 1997b.*

conversion of biomass to bioproducts, advanced agricultural systems and plant/crop engineering.

In the future, bioproducts produced from plant and crop feedstocks are likely to become functional replacements for fossil-derived products. Research is needed to advance the use of plant- and crop-based inputs in existing processing systems, to develop new or modified production systems to provide desirable plant and crop feedstocks, and to integrate these approaches into optimised manufacturing systems based solely on biomass-derived inputs.

Advanced agricultural systems enable the collection and use of increasingly detailed site-specific information with traditional farm management to better manage farming operations. These systems provide an integrated capability to improve environmental quality while enhancing economic productivity by increasing energy efficiency, optimising fertiliser and other chemical applications, and conserving soil and water resources. The technologies involved with these systems include global positioning satellites and ground systems as well as remote and in-situ electrical, magnetic, optical, chemical and biological sensors. Additional contributions are expected from advanced artificial intelligence and information networking technologies; autonomous control and robotics systems; materials that respond to soil, crop, moisture, pest and microclimate conditions; and biological and chemical methods for microbial process manipulation.

Plant/crop engineering involves improving plant productivity and utility in capturing solar energy and converting it to chemical energy. It results in increased biomass production, increased carbon fixation, improved nitrogen utilisation and recycling, and improved biomass conversion technologies. Engineering plants for end use could result in new or modified plants that can serve as biological factories for producing chemicals, enzymes, materials and fuels. Representative techniques include control of physiological processes that determine a plant's ability to grow on low nitrogen and to recycle nitrogen, manipulation of cell wall structure and assembly to produce high-strength structural wood, and insertion of single genes into target biomass species. Research is needed to better understand the metabolic pathways controlling plant productivity, to improve the efficiency of gene insertion and to further the sequencing and characterisation of gene function.

Basic Research to Support a Carbon Management Strategy[53]

Research programmes in support of an overall carbon management strategy, which includes the technologies mentioned above, must have a strong basic research component. Improvements in the fundamental understanding of biological, physical and chemical processes can lead to technologies that will help to transform energy production in a way that could sharply reduce emissions of greenhouse gases. Some of these research needs are as follows:[54]

- **Capture of CO_2, Decarbonisation Strategies, and CO_2 Storage and Utilisation:** Research areas include better gas separation; catalysts and biological processes for the decarbonisation process; solvents, membranes, molecular sieves and catalysts for capture of carbon; corrosion-resistant materials; and the environmental impacts of CO_2 sequestration in oceans and in oil and gas wells.

- **Hydrogen Development and Fuel Cells:** Expanded research is needed in catalytic, materials and membrane science and on biological, thermochemical and electrochemical hydrogen production. New and better catalysts to enable charge transfer from hydrogen at low temperatures, to improve the electrolysis of water, are also needed.

- **Enhancement of Natural Carbon Storage:** This area includes fundamental research on carbon cycling in soils and mechanisms to

---

53. *The source for this section is U.S. DOE 1997a.*
54. *Some of these research needs were mentioned in connection with specific technologies earlier in this chapter.*

enhance carbon storage in soils and oceans, and on the influence of climate change and anthropogenic emissions on the natural carbon cycle in the terrestrial biosphere and oceans.

■ **Biomass Production and Utilisation:** This area includes research on basic plant processes; development of genetic knowledge that will help in the engineering of biomass genotypes; and research on plant assembly fundamentals, soil productivity, water availability and soil carbon storage. Also needed is crosscutting research in molecular biology, biochemistry, biomimetics, bioengineering, structural biology, and environmental and social issues.

■ **Improvement of the Efficiency of Energy Production, Conversion and Utilisation:** Research areas include new and improved catalysts, improved combustion efficiency, advanced materials, better understanding of electrochemical processes, advanced sensors and controls, and advanced computational and visualisation tools.

An Example of an Energy-Technology Time Line

Because energy-producing and energy-consuming capital stock does not turn over instantaneously upon the availability of new technologies, the improvements discussed here will often be available long before their impact on greenhouse gas emissions can be discerned. One of the primary sources for this chapter (U.S. DOE 1997b) includes an illustrative time line for commercial availability of selected advanced technologies as cost-effective alternatives to conventional technologies. The time line accounts for the fact that many years will usually be required before they are mature technologies. This time line, only slightly modified, is reproduced in Figure 4.1 below. Many other technologies could be added.

Figure 4.1: Illustrative Time Line of Anticipated Energy-Technology Products[1]

2000	2005	2010
1 kWh/day refrigerator	Fuel cells for providing combined heat and power in commercial buildings	Widespread use of hybrid lighting, combining light concentrators, efficient artificial sources and fibre-optic distribution systems
Advanced turbine system for industrial cogeneration (80% total efficiency)	R-10+ windows and electrochromic windows	
Direct-reduction steelmaking	Advanced inert anode cell for aluminium production	Advanced membrane technology for chemical separation
Advanced systems for combustion of black liquor and for drying paper in pulp and paper mills	Thin strip casting in steel	Closed-cycle paper-making technology
Advanced sensors and controls for optimising industrial processes	New energy-efficient catalysts for chemical synthesis	Advanced conventional vehicle with three times the fuel economy of conventional vehicles
Direct-injection stratified-charge gasoline engine	Gasoline-electric hybrid vehicle	
Advanced heavy-duty diesel	Clean diesel for light-duty trucks and sport-utility vehicles	Real-time monitoring of water and nutrients in agricultural systems
	Cofiring of biomass with coal	
	Wind-generated electricity at 2.5 US cents per kWh	Hybrid fuel cell–advanced turbine system for power generation with 70% efficiency
Energy Efficiency	Life-extension and generation-optimisation technologies available for nuclear plants	Biofuels compete with petroleum-based transport fuels
Clean Energy		Clean coal technologies (for example, pressurised fluidised bed combustion, integrated gasification combined-cycle) increase efficiencies to 55%
Carbon Sequestration	Superconducting cables for underground transmission	
Clean Energy and Carbon Sequestration		
		Superconducting transformers and industrial motors of 200 horsepower and higher
		Injection of carbon into aquifers

1. *Likely time frames for commercial availability of technologies as cost-effective alternatives to conventional technologies. Adapted from U.S. DOE 1997b.*

2015	2020	2025	2030
Widespread production of chemicals from biomass feedstocks	Phase-change building materials with storage capacity and adaptive release rates	NEW SYSTEM: Mass-produced customised buildings with integrated envelope and equipment systems designed and sized for specific sites and climates	NEW SYSTEM: Broad-based biomass industry with new crops and feedstocks producing food, transport fuels, chemicals, materials and electricity
Hydrogen fuel-cell vehicle	NEW SYSTEM: Widespread application of industrial ecology principles to industrial systems, with linked industries and energy cascading		NEW SYSTEM: Energyplexes that integrate the fossil-fuel–based production of power, fuels and chemicals from coal, biomass and municipal wastes with nearly zero carbon emissions
Superconducting generators for utility systems		Travel demand reductions through real-time information systems	
Diesel fuels from natural gas	Production of hydrogen from solar conversion of water		
Photovoltaics for distributed and peak/sharing utility systems[2]	Widespread use of concentrating solar power technology for electricity generation	Advanced geothermal hot dry rock and magma energy systems	Utility-scale photovoltaic systems
	Simultaneous gas hydrate production and CO_2 sequestration	NEW SYSTEM: Mature hydrogen supply infrastructure enabling multiple modes of hydrogen-based transport	NEW SYSTEM: Fission reactors with proliferation resistance, high efficiency and lower costs
	Feasibility of oceanic sequestration established		
			Enhanced natural CO_2 absorption

2. *Large niche markets for distributed photovoltaic technology (such as for rooftops) are expected by 2010 – for example, in Japan. Photovoltaic technology is not expected to be used as part of central power generating stations before 2020.*

The figure serves to illustrate the expected development of individual component technologies followed by the emergence of novel, integrated systems (U.S. DOE 1997b). For example, with enhanced R&D, biofuels could be cost-competitive with other transport fuels by 2010. They could be distributed by the existing structure of service stations with only modest modifications. Later, R&D and demonstration could enable widespread production of chemicals from biomass feedstocks. Ultimately, a broad-based biomass industry could emerge with new crops, feedstocks and distribution systems producing food, transport fuels, chemicals, materials and electricity.

Both incremental technology improvements and breakthroughs in basic science are needed to advance along technology pathways such as these.

CHAPTER 5
MAXIMISING TECHNOLOGY'S CONTRIBUTION: OVERCOMING BARRIERS TO TECHNOLOGY ADOPTION

Overview

A large-scale transformation of any part of the energy system usually involves a multitude of changes, many of which are difficult to bring about. Some of them are minor but have to be made by large numbers of producers, consumers and providers of associated services. Some are major and costly, such as changes in the infrastructure that delivers energy to the consumer. Nevertheless, such transformations do occur where there is sufficient economic incentive or government intervention. During the oil crises in the 1970s, for example, high energy costs provided an incentive to move toward greater energy efficiency. Government intervention has helped make energy systems less polluting and more efficient. In some cases this has involved legislation specifying measures such as emissions and efficiency standards. In other cases, government and industry have co-operated in the development and implementation of voluntary actions.

Over the last decade, the factors that tend to impede or slow the adoption of commercial or near-commercial technologies have come to be called "market barriers"; many of these were mentioned in Chapter 3. They have been analysed in depth, and a list of barriers to market acceptance has emerged (see the text box).[55] As was the case for many of the technologies described in Chapter 3, the most important barrier

---

55. *For a fuller discussion, see IEA 1997c.*

Market Barriers and Market Failures

The most important market barrier for many technologies is that they cost more than older technologies. That is, the new technologies cannot pass the market test, given the current prices of older technologies and competing fuels. Several other specific barriers are often identified, although many of them can be quickly translated into higher costs:

- *Individuals and businesses do not always have full **information** about technologies or the ability to survey and calculate the costs of all opportunities.*
- ***Transaction costs** exist, such as that of gathering information or training workers in new maintenance techniques.*
- *Real and perceived **risks** associated with new technologies can make potential users wary of investing.*
- *There can be difficulties in obtaining **financing**.*
- ***Infrastructure investments** may be needed that are beyond the capacity of any one market actor.*
- *Technology may be efficient but **not well adapted** to users' non-energy needs and tastes.*
- *There may be **environmental barriers** to using some technologies even though they may be attractive in terms of their overall environmental impact – for example, the noise and visual effects of wind generators.*
- ***Regulations** unrelated to energy may require time-consuming procedures, evaluations and certification that can delay deployment.*
- *Slow rates of **capital stock turnover** and expansion constrain the rate at which new technologies can enter the market.*
- ***Signals about market interest** in new technology may not reach developers and marketers.*

For the most part, the above barriers reflect the normal workings of the economy. Some additional barriers involve "market failure" – the failure of market processes to allocate resources efficiently:

- **External effects**, in which the impact of producing or using energy is borne by parties other than the producers and consumers, but no compensation is made through the market. Emissions fall into this category. Internalising the costs (the "polluter pays" principle) is possible, though the exact levels of cost, damage and compensation cannot be calculated on an individual basis. Internalising costs even in a rough way can be an efficient way of encouraging market actors to avoid external effects. Unfortunately there are often powerful political obstacles to doing so. To the extent that the environmental costs of emitting carbon are not internalised in energy or technology prices, market prices can be said to be distorted.

- **Public goods**, where individuals or businesses have access to a good without having to pay for it. Because suppliers of a public good cannot collect a price from all who consume it, they tend to produce too little of it despite its value to society. Some aspects of long-term R&D are public goods, as is an increase in energy diversity of a sort that enhances a country's energy security. Information provision about new energy technologies also has some public good elements.

- **Market organisation** may not facilitate efficient decisions – for example, in the building sector, where savings from energy efficiency do not accrue to those who make the initial design and investment decisions (architects, real estate developers, landlords and so forth.).

Based on market principles, government action is called for to remove market failures or offset their effects. The fact that governments are often not willing to take strong action to do so can sometimes be the basis for other government actions that will reduce the first set of market barriers. For example, if it is too difficult for governments to internalise into market prices the full social costs of using some fuels, an alternative approach might be to help consumers reduce the transaction costs associated with the adoption of new technologies by way of programmes for information dissemination.

to using new technologies is usually higher capital cost relative to the cost of purchasing or continuing to use conventional technologies. Several other categories of barrier can exist. Some relate to one-time or indirect costs, such as the cost of disseminating and receiving information about new technologies and the transaction costs of making and implementing a purchase decision. Other important barriers are price distortions, the aversion to risk of both buyers and sellers, imperfect financial markets and organisational imperfections. The last include systems in which the technology purchase decision is made by someone other than those who would benefit from the new technology. The need for new or modified infrastructure to support new ways of producing and using energy can act as an important barrier, as can the slow pace of capital stock turnover. Overcapacity in electricity generation and low fossil-fuel prices also impede investment in new electricity generating technologies, particularly those based on renewable energy, tending to "lock them out" of the market.[56]

As noted in the text box, some of the barriers occur where markets fail to allocate resources efficiently by failing to take account of all the costs and benefits involved. These cases clearly call for actions by governments. Unfortunately, technical and political obstacles limit the ability of governments to remedy market failures at a general level, by internalising the cost of energy-related pollution into market prices, for example. In the presence of these limitations, government solutions also implement policies and programmes designed to reduce the effects of other barriers in cases where there will be a net economic benefit[57] from doing so. These solutions may take the form of tax incentives, subsidies, efficiency standards or voluntary agreements with industry, for example.

Focusing on market barriers can give the false impression that they can be easily removed by way of programmes and policies. To a great extent

---

56. *Another way to think about the issue of market barriers is to consider factors that may act as "determinants of the rate of technology adoption", such as relative technology cost, the availability of information, the level of transaction costs and the availability of infrastructure.*
57. *Broadly defined to account for external costs and benefits.*

market barriers are names given to the normal working of the economy in practice – new technologies and ways of doing things require changes in processes, attitudes and capital stock and this has to be faced. One important approach is therefore to consider processes that will improve the normal ways in which technology adoption proceeds, particularly from a systems perspective. This way of thinking about the problem is the subject of the section titled, "Overcoming Other Barriers in the End-Use Sectors".

Chapter 3 identified the most important barriers impeding the commercialisation and wide adoption of some technologies with significant potential to reduce carbon emissions by the Kyoto time frame or shortly beyond. Most of the barriers mentioned in that discussion fall into the categories of *cost, infrastructure needs, slow capital stock turnover, market organisation, other market barriers reflecting the normal workings of the market, and technical barriers requiring further R&D*. The longer-term technologies highlighted in Chapter 4 will eventually face similar barriers, and also face the *need for long-term R&D* support to bring them to market readiness. The remainder of this chapter is organised around the first five of these categories of barriers and provides examples of how they can be overcome. Chapter 6 addresses the role of government in R&D.

Overcoming Technology Cost Barriers

Cost Barriers Associated with Distorted Market Prices

A discussion of higher cost (relative to alternative technologies) as a barrier to wider use of low-carbon technologies, and how to overcome it, would be misleading if the broader economic environment were not brought into focus. Under the expected future conditions of continued economic growth and relatively low energy prices, carbon emissions will likely increase. Fossil-fuel resources have been plentiful and relatively cheap, and they are expected to remain so through the Kyoto time frame. In other words, there are powerful, market-based reasons why energy-related emissions of CO_2 in OECD countries are expected to be 30 percent higher in 2010 than they were in 1990. Powerful government policies will be needed if the growth in emissions is to be curtailed and reversed.

It is not the purpose of this report to determine the overall cost that economies should incur to meet the Kyoto commitments. Estimates of the cost of achieving domestic emissions reductions large enough to meet these commitments have ranged from as low as 10 US$ per metric ton to well above 200 US$ per metric ton. Since flexibility mechanisms and non-energy policies can also be used to meet the Kyoto commitments, the maximum cost that should be incurred in IEA domestic energy sectors is highly dependent on the expected cost and availability of emissions reductions from these other sources. Nonetheless, it is important to realise that large-scale adoption of lower-carbon energy technologies will not occur simply as a result of "no-regrets" policies or programmes that help make the energy economy more efficient and environmentally aware. To the extent that the environmental costs of emitting carbon are not internalised in energy or technology prices, this is a market failure resulting in distorted market prices. As a result, large changes to energy markets will require a changed price structure favouring low-carbon and low-emissions technologies, or regulations and subsidies that will have a similar (if less efficient) effect.[58] **Unless government policies alter the market implications of low fossil-fuel prices, it is not realistic to expect energy technology to make a great contribution toward meeting the Kyoto commitments – or further emissions reductions commitments beyond the Kyoto time frame.**

It can be noted briefly that government policies for accomplishing this task include various energy and carbon tax policies and emissions cap-and-trade systems, which send a direct price signal to the market, as well as more indirect policies such as regulations (for example,

---

58. *Countries will face different challenges in making such changes, depending on their states of development and natural resource endowments.*

emissions or efficiency standards), subsidies for cleaner fuels or technologies and removal of subsidies for higher-carbon fuels or technologies. Some of the more indirect policies tend to overlap with those used to address other, non-cost barriers.

The next sub-section discusses how other cost barriers can be addressed, and subsequent sections do the same for other categories of barriers. But it should be kept in mind that policies to lessen these barriers do not alter the fundamental market bias toward maintaining "business as usual" – which tends to cause more expensive, but lower-carbon, technologies to be "locked out" of the market.

Other Technology Cost Barriers

Not all technologies with lower carbon emissions than competing technologies would immediately be widely used if market prices were suddenly changed to reflect environmental damage costs. This is in part because of the other barriers mentioned previously – and in part because some of the more efficient and cleaner technologies would be more expensive than alternatives even under such a changed market situation. For example, manufacturing may be more difficult and costly, materials costs may be higher, or more highly skilled and expensive labour may be required for production. In such cases, the higher cost of the new technology is a barrier to wide use, even though no market failure is involved (IEA 1997c). Governments may still wish to act to help overcome this barrier, however, if there is a net economic benefit[59] to doing so.

A direct way of tackling this cost barrier is through efforts to lower new technology costs. Two mechanisms for doing this are further R&D and support for "technology learning".

Research and Development

Cost reduction is an important focus of numerous R&D programmes, such as those focused on photovoltaic cells, batteries for electric

---

59. *Broadly defined to account for external costs and benefits.*

vehicles, and fuel cells (particularly for vehicle applications). Technology advances in materials and manufacturing techniques as well as fundamental process breakthroughs are needed in some cases to significantly reduce costs. Fuel-cell prices may have to be reduced by up to one to two orders of magnitude for fuel-cell-based vehicles to be competitive in the vehicle market.

In some cases, this R&D will be undertaken by the private sector alone; in others it is of sufficient cost, duration or public interest that governments provide direct or indirect support. (This topic is discussed further in Chapter 6.) They may provide R&D funding directly to government or private-sector researchers, or may use the tax system or other mechanisms to create a favourable environment for private-sector R&D. Demonstration projects – particularly for large-scale technologies – have been identified by industry as one area where continuing government-industry partnerships are needed.

Support for Technology Learning

Experience with technology production that serves to bring down costs is referred to as "technology learning", and actions to stimulate increased cumulative production (and use) of technologies are said to drive them down the "experience curve", which is a plot of cumulative production versus unit price. [60]

There is strong evidence across industries that experience with supplying technologies reduces prices and that there is a simple, quantitative relationship between activity and price. The decrease in price that comes with cumulative production can be characterised by a "progress ratio", which indicates how quickly prices come down. A progress ratio of 80 percent means that prices are reduced by 20

---

60. *Some readers may be more familiar with the terms "learning curves" and "learning curve effect" to denote what are described here as "experience curves" and "technology learning". The Boston Consulting Group distinguishes between learning and experience curves as follows: Learning curves relate performance to one specific input to the process, such as labour hours, while experience curves relate performance to all inputs – that is, to total costs (Abell and Hammond 1979). Of interest here are experience curves, because the policy analyst can measure them by observing the market. Learning curves cannot be observed without access to plant statistics.*

percent for each doubling of cumulative production. For example, progress ratios for the semiconductor industry – which can take advantage of miniaturisation – are in the 60 to 80 percent range (Ayres and Martins 1992), while photovoltaic (PV) modules show a progress ratio of about 80 percent (IEA 2000b).

"Learning investments" are the costs over and above the corresponding cost for the least-cost alternative that are incurred to expand cumulative production. Purchasing advanced technologies before they are cost-competitive helps drive down their cost through the technology learning process. If private market actors make these investments – that is, purchase the more expensive but cleaner technology – on their own, then the market bears the full cost of the learning investments. Governments may wish to assist this process to help bring down costs faster than would otherwise be the case. They can do this by subsidising purchases, by using their own purchasing power and by using their capacity to set market rules (for example, by requiring that a certain fraction of power generation come from alternative sources, such as renewables). There may be a role for co-ordinated international procurement to provide an international boost to technology learning for specific technologies.

Government policies can stimulate learning investments on the part of market actors. Evidence shows that subsidies need not cover the entire cost differential between new technologies and their less expensive alternatives. For example, in Japan's "PV Roofs" programme, begun in 1993, each yen provided as a government subsidy stimulates 4 to 5 yen of further learning investments on the part of market actors (IEA 2000b). In Germany's 250-MW wind programme, the market provided 1.5 DM of learning investments for each DM of subsidies provided by the government (IEA 2000b).

An important tool for stimulating technology learning is the niche market. A niche market is typically a market in which a new product can compete with established alternatives because consumers are willing to pay for specific properties of the new product. At the beginning of the twentieth century, electricity found a niche market in home lighting, where it was much more efficient and cleaner than kerosene or gas light. An examples of a niche markets today is city driving, where both the nature of the trips made and concern over urban air pollution make alternative-fuel buses and other in-city vehicles applicable and valuable. Another is home-based power systems, which must be modular, applicable on a very small scale, have no emissions that cause local pollution and not require large volumes of fuel storage – a market that PV cells have been able to exploit. Niche markets such as these allow industries to gather manufacturing and operational experience and cut prices, opening up additional niche markets that may eventually provide enough experience to lead to cost-competitiveness in broader markets.

By definition, learning investments are made in technologies that are uneconomic (more expensive than alternatives, and hence unable to compete in the market) at the time that they are made, even in niche markets. They are intended to stimulate cost reductions that make the technologies more competitive outside of niche markets.

The case of PV technology illustrates how government promotion of technology learning can make use of niche markets. The cumulative production of PV modules in the world is at present only about 0.8 GW, indicating that several hundred times the current production is required before PV becomes a cost-competitive alternative in conventional applications of electricity generation technologies.[61] In the near term, the prospect of conventional application cannot be expected to drive PV production up and PV prices down. Realistic measures to make PV

---

61. *Experience curves are highly non-linear. They follow power functions, with the progress ratio defining the exponent of this power function. (Hence, they appear linear on a log-log scale.) For a technology with a progress ratio of 80 percent, this means that at cumulative production of one MW, an additional MW is needed to reduce prices by 20 percent, but that at 10 GW, another 10 GW would be needed to have the same percentage reduction. In other words, as cumulative production grows large, the time to double it naturally increases. Doubling times for new technologies may be measured in years, while for older technologies they may be measured in decades. This may give the false impression that technology learning stops after a certain cumulative volume, because the experience curve appears to be flat at that point. The lower limit on price is likely to be based on the cost of inputs (labour, capital, materials, energy) rather than due to an end to learning. But progress ratios can change over the life of a technology, and conclusions about the time required to reach cost-competitiveness must be tested for sensitivity to changes in the progress ratio.*

technology commercially competitive must rely on niche markets that put a premium on the specific characteristics of PV technology. One example of such a measure is Japan's "PV Roofs" programme, mentioned earlier. This programme uses subsidies to move the market for roof-mounted PV systems toward economic competitiveness. Experience curves are used to set up annual targets and design the subsidies. When cost-competitiveness with alternatives in this niche market is achieved, the niche market will be self-sustaining and the subsidies can be phased out. Conservative estimates are that, worldwide, this point will be reached around the middle of the next decade, but more optimistic assessments of niche market capacity in Japan forecast a self-sustaining market by 2001 (IEA 2000b).

Another approach to stimulating technology learning is exemplified by the Non-Fossil-Fuel Obligation (NFFO), which has been used in the United Kingdom to support increased use of renewable inputs to electricity generation. Under the NFFO, regional electricity companies (RECs) are required to secure specified amounts of renewable generation capacity. They contract for this capacity at above-market rates, with the difference reimbursed from a fossil-fuel levy paid by the RECs (and reflected in electricity bills). The NFFO constitutes a major governmental effort to stimulate further development of renewables technology by providing a guaranteed market. The goal of the programme is to bring the cost of renewables-based electricity down to that of electricity generated from conventional sources. The prices paid by the RECs for renewables-based electricity have indeed been falling over time (IEA 1998a). (In response to changes in the U.K. electricity market, the NFFO will be replaced with an obligation on suppliers to supply a specified percentage of electricity from renewable sources.)

Technology learning need not be the sole focus of government measures to stimulate or support technology deployment. Other actions to lower barriers and stimulate demand, such as those described in subsequent sections of this chapter, can also contribute to technology learning and consequent cost reduction.

Overcoming Infrastructure Barriers

In some cases, the use of new technology requires infrastructure investments that are beyond the capacity of any one market actor to provide. Examples include electric and alternative-fuel vehicles, which require new refuelling infrastructure, and district heating and cooling (DH&C) systems, which require distribution systems and will be more productive if they can be integrated with other energy facilities. Mass transit infrastructure is another example, as are new uses of natural gas, which may require new gas pipelines or other transport infrastructure to deliver gas to unserved areas. Additional infrastructure will also be needed to support the use of hydrogen as a fuel.

In some cases, such as refuelling, the needed infrastructure can be retrofitted into existing systems provided sufficient incentives exist for doing so, but the retrofits must generally be carried out by someone other than the technology user. This leads to a "chicken and egg" problem, wherein technology users will not adopt a new vehicle technology until the refuelling infrastructure exists and is widely enough available to make its use convenient, and potential refuelling stations will not make the necessary retrofits without a larger market of technology users.

In other cases, such as DH&C systems, the needed infrastructure is very extensive and involves the co-operation of various parties. Among other reasons, this is because the construction of DH&C systems is subject to local public-sector jurisdictions and because the attractiveness of district energy systems is partly based on their ability to use waste heat from a variety of other energy-using processes, such as electricity production and industrial processes.

Governments can invest directly in new infrastructure or provide incentives (such as tax incentives, subsidies and expedited regulatory review) for the private sector to do so. To be effective, incentives must signal a long-term commitment to new ways of delivering energy services, to provide needed investor confidence. Government efforts to increase demand for new technology – for example, by investing in alternative-fuel vehicles for its own fleet or stimulating private-sector demand – can help address the "chicken and egg" problem.

Government also has a role in facilitating infrastructure investment decisions when multiple parties are involved. For example, there is ample evidence from several countries that such a role can help overcome obstacles to investment in integrated DH&C and combined heat and power (CHP) systems. Governments can also integrate energy-efficiency considerations into broader policies affecting the end-use sectors, with a view to influencing new or replacement capital stock investments. This strategy is described in the section titled, "Overcoming Barriers Related to Market Organisation".

Overcoming Capital Stock Turnover Barriers

Many of the technologies described in Chapter 3 face the barrier of slow capital stock turnover. Society's infrastructures – transport systems, building stocks, industrial facilities and energy supply networks – change only slowly. Slow rates of turnover can also affect specific categories of equipment – industrial motors and large consumer appliances, for example. These slow rates of change retard the rate of adoption of efficient and clean energy infrastructure and equipment.

The corollary to slow stock turnover rates is long-lived infrastructure and equipment. For example, a recent IEA study (IEA 2000a) examined capital stock turnover rates and found that –

■ Thermal power plants may have an economic life of 15 to 40 years. But in the absence of policy measures, it is possible that many of today's large fossil-fuel power plants will operate well into the twenty-first century, even up to 2050.

■ Industrial equipment generally lasts at least a decade, and often as much as three decades, while structures housing industrial facilities typically last about four decades.

- Residential and commercial buildings can last for 60 to 100 years or more. Building space and water heating systems can last 15 to 25 years, while refrigerators and freezers can last 15 to 18 years.

- Passenger aircraft can last for 30 years while freight aircraft can last 35 years or more.

As a result of these long lives, missed opportunities to put efficient and cleaner stock into place when old stock is refurbished or replaced, and when new stock is constructed or purchased, can perpetuate excess emissions for a long time. Actions to influence the type of new, replacement or refurbished equipment and infrastructure put into place are therefore potentially "high-leverage" actions over the long term. Measures that can accomplish this include information programmes, equipment and building efficiency standards, emissions standards, urban planning requirements that direct development toward areas served by mass transit, and "portfolio requirements" mandating that a certain fraction of power generation be based on renewable fuels. An approach with particular promise for end-use sectors is embedding efficiency into sectoral policies (such as transport policy and housing policies), to ensure that new buildings, industrial processes and transport systems incorporate energy-efficient and cleaner technology and approaches.

An exception to long capital stock lifetimes can be found in the transport sector, where turnover of automobiles and light trucks is more rapid than for many other energy supply or consumption devices. Over the next 50 years the entire vehicle stock in IEA countries will be replaced at least two to three times. There is thus a large potential for emissions reductions, but it is unlikely that this potential can be realised in light of the dramatic slowing of the vehicle fuel-efficiency trend over the last decade. The fuel efficiency of new cars has declined by 0.6 percent per year over this period. Vehicle turnover will have little impact on transport-related CO_2 emissions unless new cars can be made substantially more efficient than the existing fleet or use lower-emitting fuels. Such an outcome only appears likely if government and industry co-operate to encourage the development and purchase of much more efficient and cleaner vehicles.

Another exception to long lifetimes is building refurbishment. Within IEA countries, buildings are refurbished much more frequently than they are replaced or than new buildings are constructed. As a result, it is important that efficiency standards and other policies that affect energy use and emissions in residential and commercial buildings be applied to building refurbishments as well as to new construction. Currently, regulations and standards for buildings apply much less frequently to refurbishments than to new construction (IEA 2000a).

In addition to influencing new, replacement and refurbished capital stock, government policies and measures may seek to increase the turnover rate of stock. This approach has not been used to a great extent; rather, the general approach has been to influence new investments. It is particularly difficult to influence the rate of capital stock turnover under conditions of low energy prices. Under these conditions, competitive pressures and market uncertainty have decreased incentives to invest in new or replacement stock in the power-generation sector. Overcapacity in electricity generation in IEA Member countries also impedes investment in new technology. In the industrial sector, major improvements generally occur only when new plants are constructed or equipment reaches the end of its economic life. And because the ratio of new cars to cars in the existing fleet is relatively small (in IEA countries) at any point in time, measures to accelerate turnover of automobiles would have relatively little effect. Measures to increase the scrappage – rather than the reselling – of the oldest cars and least-efficient vehicles could make more of a difference (IEA 2000a).

Additional research is needed into measures to accelerate stock turnover. At this point, measures that appear to have particular promise include the following (IEA 2000a):

- **Ensure that the Tax Code Does Not Constrain the Charging of Equipment Depreciation Costs Against Taxable Income.**

- **Apply Efficiency or Emissions Standards or Environmental Regulations to Existing as well as New Equipment, Vehicles, Aircraft and Power Plants.** Applying such measures to existing equipment, highway vehicles and aircraft is likely to accelerate their replacement, although the existence of many dispersed users of equipment and vehicles is likely to make enforcement difficult. Environmental regulations on emissions and waste products from power plants can make them uneconomic in their existing form and hasten fuel switching or retirements. Exemptions from regulations for old plants and the difficulties involved in gaining approval for new sites can both be strong incentives to extend the lifetimes of plants and avoid retirement.

- **Improve the Environment for Industrial Capital Investment.** General industrial capital investment plays a central role in improving energy efficiency – modern equipment and processes tend to be more energy efficient than older ones, so investment projects undertaken for non-energy reasons often also improve energy efficiency.

- **Reduce Energy Market Uncertainty.** A feature of highly competitive markets is that uncertainty about the future price of the product being supplied reduces the willingness to invest. In the electricity supply industry, uncertainty over the scope and details of market liberalisation deters investment in new and replacement capacity and motivates extension of the lives of current plants. Even after liberalisation occurs, uncertainty may continue if there is an expectation that fundamentals of the regulatory regime might be changed in the future – including aspects of the regime having to do with reducing greenhouse gas emissions. Policy makers and market regulatory authorities should as far as possible avoid creating a climate of uncertainty in moves toward liberalisation of electricity markets and in regulation of markets thereafter. Environmental policy makers and regulators should as far as possible avoid the appearance of uncertainty over the future direction of policy and the implementation of regulations.

There is little that can be done to accelerate the turnover of buildings *per se*. But improvements to building thermal integrity during building refurbishment or retrofit can be encouraged via loans or grants for example, for more efficient windows or additional insulation. They can be mandated through building codes that require that building shells and equipment be brought up to minimum standards when a certain level of rehabilitation is undertaken.

Overcoming Barriers Related to Market Organisation

Barriers related to market organisation are primarily an issue in the buildings sector, where those ultimately paying the cost of using energy (residents and businesses) often differ from those making decisions about new building thermal integrity and equipment stock (architects, designers, builders and owners). For rental property, there may be little incentive for owners, who do not pay energy bills, to upgrade building components or equipment so as to save energy.

In addition, the potential for improvement in energy efficiency in the end-use sectors is spread widely – in homes, offices, schools, hospitals, factories, communities, transport systems and so forth – and must be pursued on many fronts. Many investments to realise improved energy efficiency in these sectors are economically attractive, but are ignored because, under current energy prices, energy costs are only a small fraction of total household and business expenses for most individuals and companies. Policy actions to foster investment in improved equipment, buildings and infrastructure and to cultivate persistent attention to energy-consuming behaviour are needed.

An important approach to addressing these barriers is to integrate energy-efficiency principles into sectoral policies – housing policy, transport policy, policies affecting industry and so forth. In this way, improving energy efficiency and reducing emissions intensity over time as new capital stock turns over will become ingrained in ways of doing business in each sector.

To carry out this approach in the buildings sector, IEA Member country energy and environmental authorities will need to work closely with housing authorities to take full advantage of the ongoing modification, renewal and extension of existing buildings to ensure that investments are made in improved energy efficiency as part of these actions. Improving energy efficiency when buildings are otherwise being refurbished is less expensive than taking separate actions intended solely to improve energy efficiency. Similarly, housing authorities can be encouraged to invest in lowering costs for thermal insulation, windows and other building technologies for new and existing buildings. Finally, energy and environmental authorities can work with housing authorities to ensure that building codes and standards for existing and new buildings encourage or require energy-efficiency measures. If needed, building codes or efficiency standards can be introduced specifically for heating and cooling systems or for other building components.

In the transport sector, a number of actions are currently under way or being contemplated by transport authorities that would have the effect of shifting many fixed costs (for example, for roads) to variable costs (for example, through road pricing). Such actions are often motivated by the desire to internalise the cost of key externalities such as congestion and local pollution. While these shifts are thus not all motivated by a desire to reduce CO_2 emissions, they could have important emissions-reducing effects by reducing traffic levels, encouraging fuller use of existing trucking capacity, encouraging shifts to less carbon-intensive transit modes or encouraging the use of less energy-intensive vehicles. Energy authorities in IEA Member countries will need to work closely with transport authorities if they wish to ensure that such efforts make the maximum contribution possible to reducing CO_2 emissions.

In the manufacturing sector, the greatest increase in efficiency may come about not from direct efforts to reduce energy consumption but from pursuing other economic goals, such as improved product quality, lower capital and operating costs, and specialised product markets. The greatest influence on industrial energy efficiency may therefore result from non-energy policies such as those aimed at financial stability, industrial competitiveness or environmental cleanliness. These policies have a much greater influence on the rates of investments and types of investment in industrial facilities than can be expected from energy-efficiency policies alone. Without an economic, policy and regulatory environment conducive to investment, the market potential for improving energy efficiency is limited. But energy-efficiency policies do play a vital role in channelling investments into equipment and processes that are cleaner and more efficient than they otherwise would have been. Energy authorities in IEA Member countries will have to work closely with other relevant authorities to make the needed links between these two sets of policies.

Overcoming Other Barriers in the End-Use Sectors

As noted in the first section of this chapter, a number of barriers to the use of advanced technologies can be attributed to the normal workings of the market. Technology diffusion issues relating to the mass marketing of technologies are most important for end-use technologies (IEA 1997c). This section provides a framework for addressing the full set of barriers to greater use of efficient end-use technologies. It also describes examples of measures that have been particularly useful in addressing such barriers.

Governments and the public sector have an important role to play in transforming barriers into opportunities. At a minimum, governments can make sure that policies for energy efficiency are consistent and stable over time, so as to send reliable, long-term signals to businesses and consumers and to allow them to plan accordingly. It is also important that governments develop new mechanisms to support innovative efforts and that they use their own purchasing power in support of their goals. The procurement power of the public sector, because of its size and its ability to set an example, can be a strong force for promoting good performance, motivating technology development and increasing demand. Increased demand will contribute to lowering costs through the process of technology learning.

Business associations, governments at all levels, utilities, private interests and others have developed a variety of approaches to promote energy efficiency. Measures range from the purchasing programs of local administrations to international procurement schemes to promote the evolution of the best available technology. The general framework that has evolved is described below. Overall, efficiency policies and programmes seek to motivate participants in all sectors of the economy, including individual consumers, to rethink and improve their use of energy. They aim to reduce market barriers by providing useful information; promoting improved practices; developing more efficient products; and adopting energy-efficiency standards, targets and benchmarks. In doing so, these activities help stimulate the demand side of the energy market to adopt more energy-efficient capital stock, production processes and operating practices.

Elements of an End-Use Energy-Efficiency Policy

A recent multi-agency study (DEA/ECS/IEA 1997a) described a coherent framework for effective energy-efficiency policies. It addresses the barriers to increasing technology adoption in the end-use sectors. The elements of this framework are summarised below.

- **Establish and Maintain an Effective Market Infrastructure.** This element motivates energy users to take into account the energy consequences of their everyday decisions about behaviour, purchases, product design, research activities and investments. Actions include establishing real-cost pricing (including internalising environmental damage costs); removing subsidies and cross-subsidies; using taxes, tax subsidies and levies to support energy-efficiency policy objectives, or at least, not work against them; and securing clear rights and responsibilities of property ownership (of infrastructure, buildings and energy systems).

■ **Help Market Actors Recognise Their Best Interest and Act on It.** This element addresses the lack of investment in even those energy-efficiency technologies and measures that would provide a positive financial return over time. Actions include providing information and training to consumers and professionals, encouraging energy service companies that provide energy services or energy management for companies and public-sector institutions, and tapping special financing arrangements.

■ **Focus Market Interest on Energy Efficiency.** This element addresses market barriers relating to market organisation. As noted earlier, markets may be organised such that it is difficult to get information to all possible technology users. Energy users may be different from those who make initial equipment purchase decisions, as is often the case in the building sector). Consumers may not have adequate market power to signal their interest in energy efficiency to manufacturers. This element also addresses the existence of highly diverse sectors of the economy, such as industry, where more tailored solutions are necessary. Actions include fostering voluntary agreements, establishing and enforcing building codes and minimum energy performance standards, integrating energy efficiency into public and private procurement practices, using large-scale government purchasing to stimulate the market for advanced technology, and facilitating financing for small-scale efficiency projects by clustering them together into investment portfolios.

■ **Ensure Access to Good Technology.** This element recognises the continuing need for better technologies to be developed and, in some cases, the need for support for their deployment. It also recognises that "good" technology, from a consumer's perspective, addresses multiple needs (for better performance and so forth) and not just energy efficiency. Actions include encouraging the development and adaptation of energy-efficient technology; accelerating its diffusion through technology procurement; dealing with productive changes at the system and infrastructure levels, such as improving district heating systems; and expanding the use of CHP.

- **Develop and Maintain a Supportive Institutional Framework.** This element recognises that, because energy use is so dependent on the infrastructure that societies create for themselves, energy-efficiency principles must be part of sector policies on housing, commercial buildings, industry and transport (see the preceding section). Actions include integrating energy efficiency into sectoral policies, in collaboration with the relevant authorities, and ensuring the availability of impartial expertise.

- **Act to Ensure Continuity.** This element recognises that large-scale energy-efficiency improvements take time and require a policy approach that is clear, consistent and steadfast. Uncertainty and ambiguity in policies drain effort and resources away from meeting goals. Actions include establishing policy clarity, demonstrating leadership, implementing effective evaluation and monitoring techniques, and strengthening international collaboration.

More details are available from the study itself.

Examples of Approaches to Overcoming Barriers to Technology Adoption in the End-Use Sectors

Some illustrative examples of approaches to increasing technology adoption in the end-use sectors are described below.[62]

Approach 1: Encouraging Adaptation of Energy-Efficient Products to Increase their Overall Attractiveness

Most scenarios illustrating the potential for energy systems to change are created by calculating the effects of wider use of known, energy-efficient technologies. But to the customer, a product must be attractive for reasons beyond just energy efficiency. This is particularly

---

62. *Additional examples are available from DEA/ECS/IEA 1997b.*

the case when energy prices are low, because the anticipated cost savings are not high enough to overcome various barriers. The product must appeal to users, fit into their homes, respond to their tastes, and so forth. Industrial equipment must improve product quality or reduce materials use in addition to reducing energy consumption. Outstanding energy performance must become a standard feature of products, such that consumers benefit from it without having to make a conscious decision to do so. Outstanding energy performance must be attractive enough to producers and retailers to be considered part of good business decisions. Wider use of efficient products will result.

Example: Lighting Procurement in Sweden

Lighting in commercial areas can be improved from an energy perspective as well as ergonomically. One of the key elements is the use of high-frequency (HF) electronic ballasts. These reduce energy losses by some 15 to 20 percent and enable improved control of lighting according to occupancy or levels of daylight. But the main reasons customers purchase this equipment is that it is "flicker free" and therefore provides higher lighting quality, which improves productivity.

NUTEK[63] issued a formal procurement for HF ballasts to increase the volume demanded, thereby reducing prices and securing performance levels. The batch procured ultimately totalled 46,000 units. Almost simultaneously a demonstration programme was launched for office lighting under which requirements were specified for the fixtures, and a programme was established enabling important buyers to get more hands-on experience with new lighting designed for their specific circumstances. In the actual installations the reduction in energy use has typically been on the order of 50 percent.

As a result of this programme, the adoption of HF ballasts has multiplied by a factor of 10 to 20 in 3 years (Table 5.1). This technology

---

63. *The Swedish National Board for Industrial and Technical Development.*

has captured approximately 30 percent of the relevant market today, and interviews indicate an expected continuation of market growth.

Table 5.1

Market Development for High-Frequency Ballasts in Sweden

	1990	1991	1992	1993	1994	1995
Thousand Units Sold	10-20	50	100	220	320	600

Approach 2: Using Public Sector Actions to Set an Example, Transfer Experience and Stimulate the Market

Example: The Federal Energy Management Program (FEMP) in the United States

The federal government is the largest energy consumer in the United States. The Federal Energy Management Program (FEMP) is a customer-focused organisation providing services to other federal agencies to advance energy efficiency. It pursues its objectives by creating partnerships, leveraging resources, transferring technology and providing training and support. It sets an example by transferring federal energy experience to state and local governments and ultimately to the private sector.

The Energy Efficiency and Resource Conservation Challenge is a voluntary, government-wide commitment, within the FEMP, that uses the buying power of the federal government to support and expand markets for today's best-practice energy-efficient, renewable and water-conserving products; lower the costs of efficient products for all consumers; reduce operating costs for federal agencies, saving taxpayers' money; reduce federal energy use and greenhouse gas emissions; and provide a model for other levels of government and for corporate and institutional purchasers. The FEMP provides participants with a variety of tools and technical support to assist them in achieving Challenge goals. The FEMP publishes product recommendations that identify energy-efficiency criteria for products to be considered "best practice". Best practice products are those scoring in the upper 25 percent of energy efficiency for all similar products, or products at least 10 percent more efficient than the minimum level that meets federal standards. This system makes it easy for procurement officials to locate such products, both electronically and through catalogues and schedules.

The FEMP addresses the lack of information for consumers, business leaders and public officials on ways to implement energy-efficiency measures, particularly those incorporating advanced equipment. Often very basic information such as the amount of energy consumed for various activities is not well known. The federal government's in-house energy-efficiency measures demonstrate energy-efficiency results, cut back on waste in public expenditure and help to reduce air pollution.

Example: The Commercial Buildings Incentive Program in Canada

Canada's Commercial Building Incentive Program (CBIP) is an initiative to encourage investments in energy-efficient building design and construction. The objective is to bring about lasting changes in attitudes and design practices by demonstrating the economic and environmental benefits of energy efficiency.

The program offers building owners and developers a financial incentive to incorporate energy-efficiency features into new commercial and institutional buildings owned by the private sector, Crown corporations and municipal and provincial governments. To qualify for the incentive, a building must be at least 25 percent more energy efficient than it would be if it were constructed to meet the requirements of the Model National Energy Code for Buildings (MNECB), which is a comprehensive energy-efficiency building code that takes into account variations in regional climate conditions and energy costs.

The incentive is equal to twice the difference between the estimated annual energy costs for the approved building design and the estimated annual energy costs if the building were constructed to the MNECB standard, up to a maximum of 80,000 C$ per building and 1,000,000 C$ per company.

To support the CBIP, which began in 1998, work is under way to:

■ develop compliance verification software specifically for the CBIP;

■ work closely with utilities, other governments, and equipment and system suppliers to ensure that energy-efficiency technology is readily available for installation in commercial buildings;

■ develop technical guidelines and training courses for designers on energy-efficient design practices and on simulation of energy performance;

■ develop an energy-efficiency labelling program to increase awareness of CBIP buildings and to promote market acceptance of energy-efficient design and construction practices;

■ explore the possibility of including multi-unit residential buildings in the program.

Approach 3: Organising Procurements to Improve Technology

Many technologies suffer from a "chicken and egg" problem: insufficient demand to stimulate production, and insufficient production to expose consumers to new technologies and stimulate demand. Technology improvements can require product redesign and factory retooling, and such investments will not be made in the absence of market demand for improved products. Laboratories have developed many concepts for new technologies and products but are waiting for the market to signal its interest. In this respect, more open markets and fierce global competition are both a threat and an opportunity – a threat in that competitors might maintain market share with a lower price for an old design if a producer moves too early or too quickly with new products, and an opportunity in that producers may more easily find niche markets where the changes can be tested and paid for by customers more willing to try new solutions.

To maximise new technology's contribution to reducing greenhouse gas emissions, and to improve products, the required demand for new products must be clear in order to serve as an incentive for the developer.

Manufacturers and distributors must also calculate and manage the risk of meeting new demand with a redesigned or new technology or product. Incentives to develop and produce new products may not have to cover all costs immediately but must definitely do so in the longer run. Most manufacturers and distributors prefer to share, mitigate or hedge risks by using various partnerships and strategies. A special problem is when companies installing, operating or maintaining a new product cannot properly or easily handle the product.

The strategy of organising procurements to improve technology is based on the idea that there is potential demand for new and better products, a demand that is not organised or expressed and that manufacturers do not adequately evaluate. Procurement aims to help producers manufacture energy-efficient products that will, in this sense, correspond to consumers' needs. The impact of this type of procurement on technical change can be decisive since it pushes technologies to market that have not been exploited because of concerns about risk, cost and lack of articulated demand.

Example: Technology Procurement in Sweden

One of the best examples of procurement to transform the market for energy technology is the NUTEK programme for Energy Efficiency in Sweden. An evaluation of some 20 procurements made over the period from 1990 to 1995 shows an improvement in performance of 17 to 43 percent measured against best available product, 20 to 75 percent measured against the average available on the market and 20 to 80 percent measured against the average installed efficiency. The market response has been impressive in some cases. Sales of brine-water heat pumps increased four-fold during the procurement, from 3,000 units per year to 12,000 units per year over the space of 2 years. Of these, only 500 units were subsidised through the programme (Suvilehto and Öfverholm 1998).

The activities of the Swedish technology procurement programme have been evaluated several times. The main effects of the programme are believed to be the effects on technology development and uptake. The programme appears to stimulate additional technology purchases and further technology development by focusing attention on energy efficiency and on particular technologies and by setting a new level of performance for even those contractors that do not win contracts. But these effects are difficult to quantify with any degree of certainty and new methods are needed for their evaluation. At the same time, some findings suggest that the procurement of new technology provides a stimulus to greater energy efficiency, and that NUTEK has, through its strategic market activities, focused attention on the possibility of making energy use more efficient. These activities appear to have led to the market introduction of more energy-efficient products at an earlier stage than expected. The evaluations show, however, that the cost per unit of energy saved is relatively high for the immediate energy savings.

Example: Expanding Markets for Super-Efficient Technology in the United States

The "Golden Carrot" programme, which launched the super-efficient refrigerator initiative, demonstrated the effectiveness of partnerships between the public and private sectors to develop and promote innovative technology. Under this programme, manufacturers were invited to enter a competition to produce a more energy-efficient refrigerator. They were eligible to benefit from a 30 million US$ deployment programme to be applied to one or more refrigerator models that consumed about 30 percent less electricity than the minimum required by the 1993 U.S. federal standards. The programme was based on consumer rebates, so that "prizes" for the best technologies were dependent on sales of the winning models in the service areas of the 24 participating electric utilities. The programme's incentives both encouraged manufacturers to develop and commercialise new technology and encouraged consumers to buy the resulting product (IEA 1997c).

Building on this success, stakeholders – including utilities, community groups and government energy and environment agencies – created the Consortium for Energy Efficiency (CEE) to develop similar initiatives for other appliances and to stimulate transformation of the market for energy-using technology. The CEE employs a wide range of market transformation approaches, such as manufacturer incentives, bulk purchases and consumer education, to expand the market for super-efficient technology.

For example, in 1996, CEE and its partners were instrumental in inducing major manufacturers to produce a super-efficient refrigerator in a size category (14 cubic feet) previously overlooked by the market. A bulk procurement approach, anchored by the New York Power Authority for its customer, the New York City Housing Authority, targeted super-efficient, apartment-size refrigerators for housing authorities. Major market transformation objectives – greater product efficiency at a competitive price – were achieved.

Approach 4: Signalling Performance to Customers (Benchmarking)

Example: Energy Labels for Residences in Denmark

Mandatory energy labels for new houses and apartments in Denmark inform potential buyers of the annual energy consumption of each property and the environmental impact of that consumption. They are complemented by energy plans based on consultant reviews showing how to save energy and water and the cost involved in doing so.

This approach requires energy consultants who are well trained in energy efficiency and cost-benefit analysis. This expertise is not well developed in some countries. Measuring energy consumption requires fairly refined methods of calculation.

Example: Mandatory Energy Labels in the European Union

Domestic refrigeration appliances account for 6 percent of total annual electricity demand in the European Union. Technologically feasible and cost-effective energy-efficiency improvements are available that result in refrigeration equipment that is 50 percent more efficient than refrigerators currently in use.

To accelerate the dissemination of more energy-efficient appliances, a framework energy-labelling directive was enacted in 1992. The first energy label was for refrigerators and came into force in 1995. Since then, the average efficiency of the refrigeration equipment on the market in European Union countries has improved by 8 to 17 percent.

Approach 5: Establishing Voluntary Agreements with Industry

Example: Long-Term Agreements in the Netherlands

Long-term agreements (LTAs) are a key instrument in Dutch energy conservation and climate change policy. The agreements are legal contracts between the government and representatives from the industry and service sectors that outline broad areas of action to improve energy efficiency. They indicate the contributions expected from measures such as energy management, CHP, improvement in power generation, heat integration, and modernisation of processes. Some LTAs also specify that energy-efficiency improvements should translate into future reductions in CO_2 emissions. For its part, the government ensures some consistency and protection from new regulations. It also provides financial and technical support in exchange for voluntary participation.

The various parties to an LTA agree to pursue the same targets, though their motives may be different. The government's primary aim is reduced CO_2 emissions, while industry is driven chiefly by cost benefits and the expectation that future regulation can be prevented by active participation.

Negotiated LTAs now cover more than 90 percent of industrial primary energy consumption in the Netherlands. The target for energy-efficiency improvement for the period 1989 to 2000 is 20 percent. By the end of 1995, efficiency gains were 10 percent.

CHAPTER 6
OVERCOMING TECHNICAL BARRIERS: RESEARCH AND DEVELOPMENT AND THE ROLE OF GOVERNMENTS[64]

Governments are involved in energy-technology R&D activities to address needs that are not well provided for by private-sector activities. Left on its own, the market system does not respond adequately to certain needs that affect the public interest, such as environmental problems, especially those on the scale of global climate change. Neither does it deal effectively with certain areas of R&D, whether these relate to climate change or to other matters. If this sort of market failure is recognised, one can better understand the issues associated with one of the key messages of this report: that near-term and long-term goals should be addressed simultaneously.

Long-Term Research

The system within which energy-related R&D is conducted involves a broad range of people and institutions, including corporations of all sizes, markets, financial institutions, government R&D programmes, the education sector and non-governmental public research institutions. The different parts of the system have different roles, though there are important interactions between these roles and the processes used to implement them. At the centre of the system is the private sector: Energy technologies are primarily developed and brought into use as a result of business decisions. Government R&D activities are generally

---

64. *Sources for this chapter are the IEA Secretariat and contributions from the CERT Experts Group on R&D Priority Setting and Evaluation.*

concerned with providing basic knowledge, setting broad goals, creating and maintaining frameworks that facilitate market activities in technology development, and covering other special problems. Government R&D programmes are important, but it is the activities of business and industry that bring to fruition most of the results of such programmes.

This situation – in which the outputs of programmes financed by governments are "handed-off" to business – has a bearing on the relation between short- and long-term R&D. The initial work done by government, which "prepares the field" for business activity, typically involves R&D with long-term goals, in which it is necessary to work on problems that are distant from those associated with producing a marketable product. Business is generally not interested in carrying on such long-term R&D, primarily because it is too difficult to maintain property rights over its results and because the profits to be made from them take too long to materialise.

The question of time horizons was clarified in discussions held by the Committee on Energy Research and Technology's (CERT's) Experts Group on R&D Priority Setting and Evaluation, which met with representatives of several major industries to learn more about the business approach to energy R&D. It was clear from this discussion that business attitudes and activities differ markedly according to whether R&D involves short-, medium- or long-term time horizons. Short-term R&D (in which applications are expected within 2 years) is viewed as something that business does itself, and involvement from government is not wanted. Business is more interested in the possibility of partnerships with government on medium-term R&D (2 to 5 years), such as demonstration projects or the development of market infrastructure associated with new energy technologies. The business representatives attending the meeting indicated little interest in undertaking long-term R&D – by default, it is a responsibility of government.

Other evidence suggests that firms in industries that in the past were involved in longer-term research are now doing less long-term work. This has occurred, for instance, in utility industries in which regulatory systems have been liberalised and in oil companies that have been privatised.

In relation to the climate change problem, this specialisation of function, according to the time horizon involved in R&D activities, assigns a special responsibility to government: to ensure that enough long-term work of the right kind is done to keep the "pipeline" of potential technology improvements sufficiently full and the system within which they can be exploited operating smoothly. A large component of current government R&D can be viewed primarily as actions taken now that will shape the practical development of technologies in the future.

The need for attention to long-term research goals is greater at present because recent changes in the way governments organise their R&D programmes have had the unintended effect of shortening average programme time horizons. Rather than organising or sponsoring R&D directly, energy R&D objectives are increasingly implemented through partnerships with industry, independent or quasi-independent laboratories, and universities. Cost-shared partnerships are often instituted. In some partnerships, governments set targets for the recovery of public resources through the sharing of costs, or the amounts to be earned from work for the private sector by researchers in government laboratories.

This new emphasis on collaboration among government, industry and other organisations is efficient and productive. It increases the productivity of government R&D spending because government investment is leveraged – private spending on policy-related activities increases as a result of government initiatives – and R&D is managed better because techniques to ensure cost recovery and productivity are applied. But as an unintended side-effect, in many countries the new approach has tended to shift the funding of government activities toward shorter-term objectives and to reduce the scope of longer-term research, including projects that involve more risk than business is willing take on. To attract private-sector participation, leveraged R&D

programmes must take account of the financial needs of the corporations involved and these will generally be short-term needs. Furthermore, governments are themselves more often seeking assurances that R&D investments are in the public interest and generate a return on the government's portion of the investment. This often means that government decision-makers also favour projects in which the results can be marketed and commercialised in a short period of time (for example, 3 to 5 years). In sum, overall government R&D investment now has a nearer-term focus.

Thus not only is industry doing less long-term R&D, governmental decisions have also been contributing to shortening time horizons. The CERT's Experts Group on R&D Priority Setting and Evaluation has therefore stated that, to achieve future goals, such as preparing the field for the development of technologies that will respond to climate change needs in the longer term, it is important for governments to maintain, and in some cases, to re-establish their commitment to long-term basic research activities. Such a commitment may involve a combination of direct government funding and measures to encourage the private sector to perform more long-term R&D, where this is feasible.

Near-Term and Long-Term Goals

The above argument should not be interpreted as suggesting that government R&D activities on climate-related technologies should be limited only to long-term research. In this regard, it is useful to distinguish between R&D activities as narrowly defined and a more all-encompassing definition. The broader definition would take account of the overlap between technology development, the acceptance of new technologies by market decision makers and the public, and general energy policy. As argued in Chapter 3, most technologies that will have a significant market share around the year 2010, and will therefore have an influence on whether the Kyoto goals are met, have already passed the stage in which they depend on government R&D (as defined in the narrow sense). In some cases, it will be useful to apply some short-term R&D resources to further improvements of these technologies, but the amounts need not be large.

In the broader sense, on the other hand, government has a significant role to play in helping to develop the market acceptance of technologies already in the pipeline. In general, markets for new technologies do not take shape solely as a result of private-sector decisions. As noted in Chapter 5, governments have a role to play in reducing uncertainty as to the viability of new technologies, dealing with changes in public and private infrastructure and in general attending to necessary changes in the framework in which economic activity is carried on. In an economy in which external costs have not been internalised into market prices, governments also have a role in applying policies that will hasten the adoption of energy technologies that can contribute to dealing with environmental problems. This includes the direct or indirect provision of subsidies that will help to build markets. In addition to attracting capital investment from business, such market-building activities have important feedback effects on private-sector technology development that reduce costs and help make the new technology cost competitive. As described in Chapter 5, this process is often referred to as "technology learning".

An example of a market-building activity that has great potential is a major German programme on alternative-fuel vehicles being carried on jointly by corporations involved in vehicle manufacturing and fuel supply, in partnership with the government. It involves vehicle development, the selection of fuels to be pursued and the development of scenarios for their introduction. The organisation and support of a demonstration project of this sort can be undertaken more efficiently through the co-operation of several firms, governments and their agencies. The German programme also provides insight into how shorter-term and longer-term goals interact. From a business point of view, it involves a commitment of activities several years into the future, though it is not long-term R&D in the sense discussed above. At the same time, it will have important long-term implications. That is, if it is successful it will bring about long-term changes in the transport sector.

Another important example in which achieving both short-term and long-term goals will depend to a great extent on the actions of government is in the education and training of personnel to carry on R&D and to apply new technologies. A shortage of appropriately trained professional and technical labour is currently a major bottleneck in regard to the demonstration and application of new technologies that affect energy consumption. As one representative of industry put it, the problem is not a shortage of money for R&D, nor of ideas, but of qualified people who can carry it out. While some training is done on the job at private expense, the role of public-sector education is obviously great and this is an area that calls for increased governmental effort. The overall question of the education and training needs associated with energy-technology development has not been given adequate attention as an R&D policy issue.

Conclusion

It is clear that the challenge of making meaningful reductions in atmospheric greenhouse gas concentrations calls for a government role. The nature of the challenge calls for governments to focus on two goals at once – meeting near-term targets and taking action that will affect the capacity of governments and the private sector to continue to reduce emissions in a sustained way over the long term.

CHAPTER 7
THE WAY FORWARD: ELEMENTS OF AN ENERGY-TECHNOLOGY STRATEGY FOR REDUCING GREENHOUSE GAS EMISSIONS

As is clear from the preceding chapters, IEA Member countries face both near- and long-term climate challenges. In the near term, they have committed in the Kyoto Protocol to reduce emissions by specified amounts by the period 2008 to 2012. In the long term, they will need to reverse the growing concentration of greenhouse gases in the atmosphere, and hence will need to continue reducing emissions in a sustained way beyond the Kyoto time frame.

One of the most important messages of this report is that, given the dual nature of the challenge, a technology strategy for reducing greenhouse gas emissions must *start today*, but must *focus on both the short term and the long term at the same time*.

As individual countries develop their technology strategies for reducing greenhouse gas emissions, those wishing to maximise technology's contribution should consider the following elements:

■ Strong efforts are needed now to increase the use of efficient and clean technologies that are already commercial or near-commercial – both for energy supply and energy end use. Many available or almost available technologies could reduce emissions substantially if they were widely used.

These efforts should include technology- and sector-specific measures. But these measures are likely to have little effect without reinforcement from measures that create economy-wide price signals. Both sets of measures are indispensable.

■ At the same time, intensified efforts are needed in longer-term R&D to develop and commercialise advanced technologies, particularly those offering fundamentally different ways of providing and using energy services with greatly reduced emissions. As energy market reforms and increasing global competition drive private-sector R&D toward a concentration on the shorter term, the government's role in supporting long-term R&D becomes increasingly important. Industry has also noted the continuing need for government support for demonstration projects.

Longer-term R&D involves both basic research into enabling sciences and long-term support for technology development and commercialisation. Because of the long lead times for developing and commercialising new technology, these R&D investments must be steady and maintained.

Fundamentally new ways of producing and using energy services will require not only new technologies but also new infrastructure. Partnerships between government and the private sector will be vital to the development and demonstration of new technologies and to the creation of infrastructure to support their wide use.

■ Policies and measures to support the use of efficient and clean energy technologies must also be maintained over the long term, for two reasons:

– To provide the stability needed to foster increased capital investment. Only increased investment in new and replacement buildings, power plants and equipment will provide the capital stock turnover and expansion that creates opportunities for using more efficient and cleaner technologies and systems. Uncertainty deters such investment.

– To support wide use of longer-term technologies. Without policies that create price signals and other policies and measures to support their use, the technologies that are commercialised 20 years from now will face the same deployment hurdles as today's advanced but little-used technologies.

As efficient and cleaner technologies become competitive, it may be possible to phase out supporting policies and measures for specific technologies – but the general effort to increase the use of advanced technologies should continue.

■ Because of its large, concentrated emissions sources, the electricity supply sector is often viewed as easier to tackle than the end-use sectors in terms of emissions reduction. But efficient and clean end-use technologies have a broad impact throughout the energy economy, reducing the use of both direct fuels and electricity. Their rapid deployment could make a substantial contribution to emissions reduction. Both supply and end-use technologies should be addressed.

■ In deregulating and otherwise liberalising energy markets, care should be taken to avoid unintended side effects that run counter to climate change policy goals. For example, market deregulation efforts should avoid creating incentives to continue using, or to invest anew, in less efficient and less clean energy supply technologies. In addition, policy makers should be aware of the potential for "technology lock-out", wherein none but the cheapest technology (for example, natural gas combined cycle plants) can compete successfully in the market. Measures may be needed to support investment in alternative technologies that are more efficient or use cleaner fuels, such as renewable energy technologies and combined heat and power systems.

■ Governments have a potentially important role to play in creating and stimulating markets for new technology and in supporting "technology learning". They can do this by direct investment (through their own purchasing and through subsidies for the purchases of others) and by setting market rules (for example, efficiency standards or regulations that require that a certain fraction of electricity generation come from certain sources, such as renewable fuels). Investing in alternative-fuel vehicle fleets or advanced building shell components for governments' own use can help stimulate the technology learning process as well as the development of supporting infrastructure. Providing guaranteed markets for advanced products or organising small purchasers into larger groups can stimulate technology advancement and use. Governments may wish to consider co-ordinating their procurement internationally.

■ Once in place, power plants, buildings, transport systems and other parts of society's infrastructure are likely to be in place for a long time. Missed opportunities to use clean and efficient technology when infrastructure is expanded, replaced or refurbished can lock in higher emissions for decades to come. Measures to influence the choice of long-lived capital stock are therefore potentially high-leverage actions. Such measures may include building codes for new buildings and refurbishments; efficiency standards for equipment, vehicles and power plants; power plant emissions standards; urban planning requirements that emphasise mass transit; and measure to improve the climate for industrial capital investment. Energy authorities will need to work closely with housing and transport authorities, and with those setting policy for industries and businesses, to ensure that energy efficiency is ingrained into ways of doing business.

■ The high-leverage actions to influence long-lived capital stock take on even more importance in developing countries and in countries with economies in transition. Almost two-thirds of the projected world investment in new power-generating facilities between 1995 and 2020 is expected to occur outside the OECD (whether measured in terms of new capacity or new capital investment) (IEA 1998b). Large fractions of the building stock and district-heating infrastructure in Eastern Europe will require refurbishment in the coming decades. Co-operation between IEA Member countries and non-IEA countries could help the latter avoid locking themselves into carbon-intensive pathways over the next century.

CHAPTER 8
THE ROLE OF THE
INTERNATIONAL ENERGY
AGENCY

The IEA serves as a forum for discussion of energy matters among its Member countries and, increasingly, between Member and non-Member countries (NMCs). Its role derives from its convening power – its ability to bring countries together to share experiences and work together toward common goals – and its analyses, as well as its ability to engage the private sector. All three mechanisms can be used to support IEA Member countries in their pursuit of vigorous technology strategies to reduce greenhouse gas emissions.

Existing Co-Operation on Technology

The primary mechanism for co-operation on technology matters thus far has been the IEA legal frameworks known as "Implementing Agreements".[65] The work performed under the Implementing Agreements is overseen by three Working Parties – covering Renewable Energy Technologies, Fossil Fuel Technologies and End-Use Technologies – and a Fusion Power Co-ordinating Committee. Further guidance is offered by three Experts Groups, on Electricity, on Oil and Gas, and on R&D Priority-Setting and Evaluation. The entire technology programme is overseen by the Committee on Energy Research and Technology (CERT), which is composed of senior government officials from energy, technology or research ministries.

---

65. *The term "Implementing Agreement" refers merely to an agreement to implement the IEA's International Energy Programme, but thus far such Agreements have been used solely for co-operation on technology R&D and on operation of information centres.*

The Implementing Agreements provide valuable frameworks for co-operation on R&D tasks, for information exchange between governments and researchers, and for co-operation on information dissemination. They have not yet been used to a great extent to support technology deployment directly, although a pilot project is under way to evaluate an effort to deploy good practices for fossil power plant operation in China. In addition, some Implementing Agreements are developing specific tasks focused on deploying their respective technologies in NMCs. Efforts are currently under way within the IEA Secretariat to expand co-operation on deployment of renewable energy technologies through the Renewable Energy Working Party and to motivate manufacturers to incorporate technology to reduce "leaking electricity" into equipment and appliances.

The IEA and Non-Member Countries

From the Agency's creation in 1974, its Member countries have been "desiring to promote co-operative relations with oil producing countries and with other [i.e., non-Member] oil consuming countries, including those of the developing world".[66] Fostering energy-technology innovation is a central part of the IEA's work, and the *Agreement on an International Energy Program,* which created the IEA, called for Member countries to "keep under review the prospects for co-operation with... [non-Member] countries on... research and development."[67]

Recently, the IEA has encouraged NMCs to participate in its co-operative programme on energy technology. Since 1991, IEA work with NMCs has expanded significantly through participation by selected NMCs in various Implementing Agreements. Such co-operation has been extended to NMCs whose economies are in transition, such as Russia and Poland, and countries with potentially large energy sectors, such as China and Korea. This co-operation appears to have been mutually beneficial and plans for expansion to other NMCs, such as India, are now among the priorities of the Agency.

---

66. *Agreement on an International Energy Program, Preamble, Nov. 19, 1974.*
67. *Chapter VIII, Relations with Producer Countries and with Other Consumer Countries, Article 47.*

In addition, through its work with the NMCs, the IEA has been able to provide its Member governments with information on energy policy developments. Energy Policy Surveys by the IEA have covered the Middle East, Central and Eastern Europe, Russia, Asia (China and India), Latin America and Africa.

The IEA has focused increased attention on the development of safer, more efficient technologies for global environmental protection, in particular to address climate change, and economic growth. With this shift, the IEA increasingly emphasises widespread deployment of more economical and environmentally benign technologies. Such deployment is especially important in the case of the NMCs, which are projected to make significant contributions to atmospheric buildup of greenhouse gases from energy production and use. Securing NMC commitment to technology development and deployment through participation in IEA activities is a continuing IEA priority.

The CERT and its Working Parties are exploring several opportunities for further involvement with the NMCs. For countries such as China and India, with a science and technology infrastructure comparable to that in many IEA Member countries, there is ample opportunity for mutually beneficial co-operation in virtually all the areas covered by the Implementing Agreements.

A serious constraint may be finding financial support for covering the costs of NMC participation. For some NMCs to be effective participants (and not merely the beneficiaries of information exchange), one possible approach could be to secure third-party funding through a donor, such as the Asian Development Bank, the World Bank or the United Nations.

Potential News Areas for Co-Operation and Analysis

In addition to the current and proposed modes of co-operation, there may be additional ways in which IEA Member countries can co-operate that would support their efforts to develop and implement technology strategies for reducing greenhouse gas emissions. The IEA's *convening power* could be used to facilitate co-operation in numerous areas, including the following:

■ technology road-mapping exercises, also known as "visioning" or "foresight" exercises, to identify long-term R&D priorities for potential co-operation through the Implementing Agreements or other mechanisms, and to share experiences with various ways of conducting such exercises;

■ harmonisation of codes and standards, where appropriate;

■ working with industry – for example, the current initiative to reduce "leaking electricity" losses from electronic equipment;

■ exchange of information – on a regular basis – on experiences with measures to overcome barriers to wide use of advanced technology;

■ exchange of information on experiences with measures to stimulate increased private-sector R&D;

■ exchange of information on experiences in working with industry on co-operative government-industry R&D;

■ co-operation on expanded efforts to transfer best technology practices to developing countries;

■ expanded co-operation on technology R&D in areas such as electricity technologies and nuclear fission technology, where there is currently little or no coverage by Implementing Agreements;

■ possible co-operative efforts to support technology learning, such as co-ordinated procurement efforts across some or all IEA Member countries.

The particular value of IEA analysis products is their cross-country perspective, which supports Member countries' efforts to share information and learn from one another. IEA *analysis* in support of Member countries' efforts to develop a technology strategy could focus on areas such as the following:

- assessing Member country experiences with
  - measures to overcome barriers to wide use of advanced technology
  - measures to stimulate increased private-sector R&D
  - working with industry on co-operative government-industry R&D;

- gathering and analysing additional data on energy capital stock turnover and conducting additional research into Member country experiences with measures to accelerate stock turnover;

- assessing in detail the contributions that various technologies might make to reduce greenhouse gas emissions across IEA Member countries.

The IEA can also help Member countries in broad efforts to *engage the private sector* in the search for the most cost-effective ways to induce accelerated development and adoption of technologies.

REFERENCES

Abell, D.F. and J.S. Hammond. 1979. "Cost Dynamics: Scale and Experience Effects". In *Strategic Planning: Problems and Analytical Approaches*. Englewood Cliffs, NJ, USA: Prentice Hall.

Automotive Fuels Information Service (AFIS). 1999. *Automotive Fuels for the Future: The Search for Alternatives*. Paris: OECD/IEA.

Ayres, R.U. and K. Martins. 1992. "Experience and the Life Cycle: Some Analytic Implications". *Technovation*, Vol. 12, pp. 465-486.

Bouma, J. 1994. "Residential and Commercial Heat Pumps". *Proceedings: Energy Technologies to Reduce CO_2 Emissions in Europe: Prospects, Competition, Synergy*. Conference held in Petten, the Netherlands, 11-12 April 1994. Paris: OECD/IEA.

Broderick, James R. and Alex Moore. 2000. "Conquering Corrosion: Developing New Alloys for Furnaces". *ASHRAE Journal* Vol. 42, No. 4.

Danish Energy Agency, Energy Charter Secretariat, and International Energy Agency (DEA/ECS/IEA). 1997a. *Energy Efficiency Initiative, Volume I: Energy Policy Analysis*. Paris: OECD/IEA.

Danish Energy Agency, Energy Charter Secretariat, and International Energy Agency (DEA/ECS/IEA). 1997b. *Energy Efficiency Initiative, Volume II: Country Profiles and Case Studies*. Paris: OECD/IEA.

European Commission (EC). 1997a. *A Community Strategy to Promote Combined Heat and Power (CHP) and to Dismantle Barriers to its Development*. COM(97) 514 (final). Brussels: EC.

European Commission (EC). 1997b. *Energy Technology: The Next Steps. Summary Findings from the ATLAS Project*. DG XVII-97/011. Brussels: EC.

Fusion Power Co-ordinating Committee. See Appendix A.

Gas Research Institute (GRI). 1996. "Anatomy of a Modern Gas Furnace". *GATC Focus*, May.

Implementing Agreements. See Appendix A.

Intergovernmental Panel on Climate Change (IPCC). 1996. "IPPC Technical Paper I: Technologies, Policies and Measures for Mitigating Climate Change". Geneva: World Meteorological Organisation and United Nations Environment Programme.

Interlaboratory Working Group (IWG). 1997. *Scenarios of U.S. Carbon Emissions: Potential Impacts of Energy Technologies by 2010 and Beyond.* LBNL/40533 and ORNL/CON-444. Berkeley, California: Lawrence Berkeley National Laboratory; Oak Ridge, Tennessee: Oak Ridge National Laboratory.

International Energy Agency (IEA). 1994. *IEA/OECD Scoping Study: Energy and Environmental Technologies to Respond to Climate Change Concerns.* Paris: IEA/OECD.

International Energy Agency (IEA). 1995. *The IEA Natural Gas Security Study.* Paris: IEA/OECD.

International Energy Agency (IEA). 1997a. *Electric Technologies: Bridge to the 21st Century and a Sustainable Energy Future.* Report of the First IEA Workshop on the Role of Electric Technologies in Mitigating Greenhouse Gas Emissions, Paris, September 1997. Paris: IEA/OECD.

International Energy Agency (IEA). 1997b. *Energy Technologies for the 21st Century.* Paris: IEA/OECD.

International Energy Agency (IEA). 1997c. *Enhancing the Market Deployment of Energy Technology: A Survey of Eight Technologies.* Paris: IEA/OECD.

International Energy Agency (IEA). 1998a. *Energy Policies of IEA Countries: The United Kingdom, 1998 Review.* Paris: IEA/OECD.

International Energy Agency (IEA). 1998b. *World Energy Outlook, 1998 Edition.* Paris: IEA/OECD.

International Energy Agency (IEA). 1998c. Unpublished papers from the Second IEA Workshop on the Role of Electric Technologies in

Mitigating Greenhouse Gas Emissions. Sponsored by the Edison Electric Institute, Unipede/Eurelectric and the Japanese Federation of Electric Power Companies (FEPC). Paris, October 1998.

International Energy Agency (IEA). 2000a (forthcoming). *Energy Capital Stock Turnover: A Critical Element in Reducing Future Carbon Dioxide Emissions*. Paris: IEA/OECD.

International Energy Agency. 2000b. *Experience Curves for Energy Technology Policy*. Paris: IEA/OECD.

Mintzer 1999. "Renewable Energy Technologies". Unpublished paper prepared for the International Energy Agency by Irving Mintzer.

President's Committee of Advisors on Science and Technology (PCAST). 1997. *Report to the President on Federal Energy Research and Development for the Challenges of the Twenty-First Century*. U.S. PCAST Panel on Energy Research and Development. Washington: Executive Office of the President.

Scott, David and Per-Axel Nilsson. 1999. *Competitiveness of Future Coal-Fired Units in Different Countries*. London: IEA Coal Research, The Clean Coal Centre.

Suvilehto, H-M and E. Öfverholm. 1998. "Swedish Procurement and Market Activities – Different Design Solutions on Different Markets". In *Energy Efficiency in a Competitive Environment: Proceedings of the 1988 ACEEE Summer Study on Energy Efficiency in Buildings*. Vol. 7, pp. 311-322. Washington: American Council for an Energy Efficient Economy.

U.S. Department of Energy (DOE). 1995. *Sustainable Energy Strategy: Clean and Secure Energy for a Competitive Economy*. Washington: U.S. DOE.

U.S. Department of Energy (DOE). 1997a. *Carbon Management: Assessment of Fundamental Research Needs*. Washington: U.S. DOE, Office of Energy Research.

U.S. Department of Energy (U.S. DOE). 1997b. *Technology Options to Reduce U.S. Greenhouse Gas Emissions*. Washington: U.S. DOE.

U.S. Department of Energy, Energy Information Administration (DOE/EIA). 1998. *Impacts of the Kyoto Protocol on U.S. Energy Markets and Economic Activity*. SR/OIAF/98-03. Washington: U.S. DOE.

Working Party on Fossil Fuels. See Appendix A.

World Energy Council (WEC). 1995. *Energy Efficiency Improvement Utilising High Technology: An Assessment of Energy Use in Industry and Buildings*. Report and Case Studies. London: WEC.

APPENDIX A: IMPLEMENTING AGREEMENTS AND CERT SUBSIDIARY BODIES CONTRIBUTING TO THIS REPORT

Technical contributions to this report were provided by experts involved in the IEA's collaborative research and development activities, which are conducted under a legal framework referred to as an "Implementing Agreement." Contributions were also provided by experts in the subsidiary bodies of the IEA Committee on Energy Research and Technology. Substantial contributions were made by the Implementing Agreements and subsidiary bodies listed below, which can be contacted for further information. (Please check with the IEA Secretariat for current contact information.)

CERT Subsidiary Bodies

- Experts Group on R&D Priority-Setting and Evaluation

- Fusion Power Co-ordinating Committee

- Working Party on Fossil Fuels

Implementing Agreements

- Advanced Fuel Cells

- Assessing the Impacts of High-Temperature Superconductivity on the Electric Power Sector

- District Heating and Cooling

- Energy Conservation in the Pulp and Paper Industry

- Energy Conservation in Buildings and Community Systems

- Energy Conservation through Energy Storage

- Energy Technology Systems Analysis Programme (ETSAP)

- Geothermal Energy

- Greenhouse Gases R&D Programme

- Heat Pumping Technologies

- Process Integration Technologies

- Implementing Agreement on Solar Power and Chemical Energy Systems (SolarPACES)

- Implementing Agreement on Wind Turbine Systems

APPENDIX B: ABBREVIATIONS

AC	alternating current
ATS	advanced turbine system
BIG/GT	biomass-integrated gasification/gas turbine
CBIP	Commercial Buildings Incentive Program (Canada)
CCGT	combined-cycle gas turbine
CEE	Consortium for Energy Efficiency (United States)
CERT	Committee on Energy Research and Technology (IEA)
CFL	compact fluorescent lamp
CHP	combined heat and power
CO_2	carbon dioxide
DC	direct current
DH	district heating
DH&C	district heating and cooling
EIA	Energy Information Administration (United States)
ETBE	ethyl-tertiary-butyl ether
ETSAP	Energy Technology Systems Analysis Programme
EU	European Union
FBC	fluidised bed combustion
FEMP	Federal Energy Management Program (United States)
GW	gigawatt (1 billion watts)
HF	high frequency
HHV	higher heating value
HVAC	heating, ventilation and air conditioning
IEA	International Energy Agency
IGCC	integrated gasification combined-cycle
ITER	International Thermonuclear Experimental Reactor
kW	kilowatt (1,000 watts)
kWh	kilowatt-hour
LHV	lower heating value
LTA	long-term agreement
LWR	light-water reactor
MCFC	molten carbonate fuel cell

MNECB Model National Energy Code for Buildings (Canada)
MPa   Megapascal (1 million pascals)
MTBE   methyl-tertiary-butyl ether
MW   megawatt (1 million watts)
NFFO   Non-Fossil-Fuel Obligation (United Kingdom)
NGCC   natural gas combined-cycle
NMC   non-Member country (of the IEA)
NO_x oxides of nitrogen
OECD   Organisation for Economic Co-operation and Development
PAFC   phosphoric acid fuel cell
PEMFC proton exchange membrane fuel cell
PF   pulverised fuel
ppm   parts per million
PV   photovoltaic
R&D   research and development
REC   regional electricity company (United Kingdom)
SOFC   solid oxide fuel cell

APPENDIX C: SUMMARY PAPER FOR ENERGY MINISTERS

THE TECHNOLOGY RESPONSE TO CLIMATE CHANGE – A CALL FOR ACTION

The IEA's Committee on Energy Research and Technology (CERT) has been considering how science and technology can be mobilised to help IEA Member countries meet the commitments entered into at Kyoto. The results of this study, which reflects inputs from CERT's subsidiary bodies and Implementing Agreements, are set out in the report, "The Role of Technology in Reducing Energy-Related Greenhouse Gas Emissions". The CERT believes there are a number of important messages flowing from this report which Ministers should consider as a matter of urgency.

Recognising the Nature of the Problem

New deployment policies are necessary

First, it is vital that ministers do not underestimate the scale of the effort that will be needed to meet the Kyoto commitment. In the absence of any additional action, greenhouse gas emissions will rise more than 20% above 1990 levels; i.e. to meet the Kyoto target the "real" gap to be bridged in 2008-2012 is on the order of 30%. Only today's commercial and near-commercial technologies will contribute to reducing emissions in this time frame. While these technologies make it possible to meet the Kyoto targets, under business-as-usual conditions, they will not be deployed on a sufficient scale for the targets to be met. Current trends in technology adoption are insufficient to meet the Kyoto targets. Policies and measures to accelerate technology deployment will be required.

Immediate and sustained action is required

2010 may seem a long way off, but it will take time to implement new policies and measures. Also, due to the time required for technology introduction and capital stock turnover, many promising technologies will not significantly reduce emissions before 2012. Immediate and sustained action will be required if technology is to play an important role in reducing emissions by 2012 and beyond.

Governments should increase long-term research and development

Long-term emission reductions will continue to be important since the social cost of carbon is expected to increase after 2012. The development and use of new technologies should play an even stronger role in meeting this long-term challenge. This requires an increased R&D effort now to lay the foundations. In most IEA countries, government-funded energy R&D has been declining, sometimes steeply, for at least a decade. The long-term R&D that will provide tomorrow's advanced technologies is losing priority. Governments should increase investment in long-term R&D. It is also important to provide a good environment for private-sector R&D.

The Contribution of Technologies

Many technologies can reduce CO_2 emissions

A range of technologies can be deployed to reduce CO_2 emissions. There is no single technology solution; every country will need to make a choice based on its own unique circumstances and conditions. Government support for these technologies, allowing for national circumstances, should be included in plans to reduce carbon emissions. For example, reducing CO_2 emissions from power generation can be achieved with natural gas combined cycle technology and by extending the life of nuclear plants. Clean-coal technologies will significantly improve the environmental performance of plant where coal is the fuel of choice. Biomass and wind are important options. Photovoltaic power is a short-term option in certain regions and will be more widely used in the long term. Fuel cells provide a longer-term option for stationary generation. In the transport sector, more efficient conventional vehicles can provide significant near-term savings while electric, hybrid and fuel cell vehicles have much larger long-term potential. For industry there is a wide range of technologies (process integration, motors and drives, separation processes, electro-technologies) giving substantial energy savings. In buildings, advances in heating, ventilation, air conditioning, glazing, lighting and insulation offer major savings in emissions. Technologies to exploit an integrated approach to production and use of energy and cross-sectoral technologies (advanced gas turbines, combined-heat-and-power, sensors and controls, power electronics) can make a major contribution. In the longer term, the capture and sequestration of CO_2 may prove significant.

The Limitations

Price distortions create barriers to the deployment of clean energy technology

Climate friendly technologies are not being deployed at a sufficient rate or in sufficient amounts to allow IEA countries to meet their targets. The main reason for this is that these are generally more expensive than conventional technologies. The current low price of energy exacerbates the problem. This price disadvantage, which is further increased by subsidies given to fossil fuels and the absence of policies to internalise the social cost of carbon emissions, is a barrier to the deployment of climate-friendly technologies. Technology deployment policies can help overcome price barriers since they encourage "technology learning". These "learning investments" will be repaid with more competitive low-carbon technologies and new cost-effective solutions to our climate problem.

Non-price barriers can also impede progress

It should also be noted that, even when significant technological advances are made, these do not always result in corresponding reductions in greenhouse gases. This may be because the improvements are in relatively small sectors or start from a small base. Renewable technologies have advanced rapidly and will grow over the next decade. However, even with rapid growth they will remain a relatively small fraction of overall power generation by 2012. Their contribution will be much more significant after 2012. Societal and behavioural factors can also constrain progress, especially when the benefits of technology are taken in ways that do not reduce energy use. For example, while there have been major advances in the engine efficiency of cars, the consumer preference for larger, heavier vehicles has completely negated these, and emissions from transport sector have risen considerably.

What Needs to be Done?

The way forward

Governments can and should play a role in transforming the current problems facing low-carbon technologies into opportunities. This change can be encouraged through procurement programmes, market stimulation measures, voluntary agreements and information programmes. However, such measures are unlikely to have sufficient impact if not reinforced by price signals or other ways to encourage investments in low-carbon technologies. Governments should reverse the decline in long-term R&D investment and intensify technology deployment policies and actions. Collaboration through the IEA enhances value-for-money in R&D, deployment policies and the promotion of technology use in Member and non-Member countries. Fossil-fuel subsidies should be removed and market incentives to reduce carbon emissions should be established.

Order Form

OECD BONN OFFICE

c/o DVG mbh (OECD)
Birkenmaarstrasse 8
D-53340 Meckenheim, Germany
Tel: (+49-2225) 926 166
Fax: (+49-2225) 926 169
E-mail: oecd@dvg.dsb.net
Internet: www.oecd.org/bonn

OECD MEXICO CENTRE

Edificio INFOTEC
Av. Presidente Mazarik 526
Colonia: Polanco
C.P. 11560 - Mexico D.F.
Tel: (+52-5) 280 12 09
Fax: (+52-5) 280 04 80
E-mail: mexico.contact@oecd.org
Internet: www.rtn.net.mx/ocde

OECD CENTRES

*Please send your order
by mail, fax, or by e-mail
to your nearest
OECD Centre*

OECD TOKYO CENTRE

Landic Akasaka Building
2-3-4 Akasaka, Minato-ku
Tokyo 107-0052, Japan
Tel: (+81-3) 3586 2016
Fax: (+81-3) 3584 7929
E-mail: center@oecdtokyo.org
Internet: www.oecdtokyo.org

OECD WASHINGTON CENTER

2001 L Street NW, Suite 650
Washington, D.C., 20036-4922, US
Tel: (+1-202) 785-6323
Toll-free number for orders:
(+1-800) 456-6323
Fax: (+1-202) 785-0350
E-mail: washington.contact@oecd.org
Internet: www.oecdwash.org

I would like to order the following publications

PUBLICATIONS	ISBN	QTY	PRICE*	TOTAL
☐ Energy Technology and Climate Change – A Call to Action	92-64-18563-1		$75	
☐ Energy Labels and Standards	92-64-17691-8		$100	
☐ Experience Curves for Energy Technology Policy	92-64-17650-0		$80	
☐ World Energy Outlook 2000	92-64-18513-5		$150	
☐ The Link between Energy and Human Activity	92-64-15690-9		$16	
☐ Electricity Market Reform – An IEA Handbook	92-64-16187-2		$50	
☐ Nuclear Power: Sustainability, Climate Change and Competition	92-64-16954-7		$60	
☐ Automotive Fuels for the Future – The Search for Alternatives	92-64-16960-1		$100	
			TOTAL	

*Postage and packing fees will be added to each order.

DELIVERY DETAILS

Name _____ Organisation _____

Address _____

Country _____ Postcode _____

Telephone _____ Fax _____

PAYMENT DETAILS

☐ I enclose a cheque payable to IEA Publications for the sum of US$ _____ or FF _____

☐ Please debit my credit card (tick choice). ☐ Access/Mastercard ☐ Diners ☐ VISA ☐ AMEX

Card no: ⌞_ _ _ _ _ _ _ _ _ _ _ _ _ _ _⌟

Expiry date: ⌞_ _ _ _ _ _⌟   Signature:

IEA PUBLICATIONS, 9 rue de la Fédération, 75739 PARIS Cedex 15
Printed in France by Louis Jean
(61 00 31 I P) ISBN 92-64-18563-1 2000